A Quick Guide to Book-On-Demand Printing
Revised Edition

Novels by Roger MacBride Allen

The Torch of Honor
Rogue Powers
(above reissued in combined volume as *Allies & Aliens*)

Orphan of Creation
The Modular Man

Farside Cannon
The Ring of Charon
The Shattered Sphere

The Game of Worlds

The Depths of Time
The Ocean of Years

Supernova (with Eric Kotoni)

The War Machine (with David Drake)

A trilogy of Asimovian Robot Novels:
Caliban
Inferno
Utopia

The Corellian Trilogy of Star Wars Novels:
Ambush at Corellia
Assault at Selonia
Showdown at Centerpoint

A Quick Guide to Book-On-Demand Printing

Revised Edition

by Roger MacBride Allen

"Freedom of the press is guaranteed only to those who own one."

—A.J. Liebling

A Quick Guide to Book–On–Demand Printing,Revised Edition

Version 3.0
ISBN 0-9709711-8-4

this text last revised: August 14, 2002
for the latest updates to the book visit
www.foxacre.com/bookpage/bodupdt.htm

Takoma Park, Maryland—Ewing, New Jersey—Leipzig, Germany
www.foxacre.com

Dedication

To Van and Ellie Seagraves
Who Got Me Started in
Do-It-Yourself Publishing

Contents

Preface to the Revised Edition* .... *x

Chapter One
Book on Demand What, How, When, Why—And Who?* *1
1. Introduction 1
2. The Economics of Book-On-Demand Printing 2
3. Purposes of This Guide 5
4. How This Guide Got Here and Where It's Going 7
5. Sources for Information In This Guide 9

Chapter Two
Publishing Realities and Publishing Choices* *12
1. Who Should Consider Book-On-Demand Printing 12
2. Who SHOULDN'T Consider Book-On-Demand Printing 13
3. Is Doing Book-On-Demand Worthwhile? 14
4. Publishers, Conventional and Otherwise 16
5. Conventional Publication 18
6. Vanity Press 23
7. Private Publishing 25
8. Conventional Self-Publishing 25
9. Third-Party Book-On-Demand Publishing Services 27
10. Web-Based Electronic Books 27
11. Proprietary-Format E-Books 29
12. Self-Published Book-On-Demand 32

Chapter Three
Getting Started* *34
1. The Basic Tasks 34
2. Equipment for Printing and Paper Handling 40

Monochrome (Black) Laser Printers 40
Color Laser Printers 45
Ink-Jet Printers 46
Paper Cutters (1) For The Initial Cut 48
Paper Cutters (2) For The Final Trim 52
Scoring Tools & Techniques 54
Varnishes and Laminators 58

Chapter Four
Page Design and Printing 64
1. Some Quick Notes on Typography 64
2. Choosing the Proper Software Tools 66
3. Paper, Paper Grain, and Paper Curl 69
4. One-Up vs. Two-Up Printing 75
5. Twin Two-Up Printing 75
6. Duplex Printing 77
7. Page Imposition 79
8. Using Custom Paper Sizes 84
To Avoid Imposition and Cutting 84
9. PostScript, Acrobat, and Page Imposition 85
10. Adding Sheets From Other Sources 91

Chapter Five
Printing Book Covers 92
1. Cover Design 92
2. ISBNs and Barcodes 93
3. Using a Commercial Printer 94
4. Printing Your Own Covers 96
5. Paper Stocks Suitable for Covers 97
6. Cover Size and Layout 101

Chapter Six
Hand Binding and "Office" Thermal Binders 105
2. Short-Run Office Thermal Binding Systems 108
3. The Evans Do-It-Yourself Thermal Binding Systems 112
4. Using Your Binder For Instant Paperback Repair 115

5. Cold Glue Binding 117
6. Side-Stapling 121
7. Gigabooks 121

Chapter Seven
Small Binding Machines 124
1. Automated Processes and Hand-Labor 124
2. Binding Strip Machines 127
3. Other Small Binding Machines 130

Chapter Eight
Production Binding Machines
and Specialty Large-Scale Machines 138
1. General Operating Principles
of Most Small Commercial Binding Machines 138
2. Table-top Binding Machines 146
3. Floor-Model Binding Machines 150
4. Commercial Cold-Glue Binding
and Lay-Flat Binding Machines 153
5. Specialty and Large-Scale
On-Demand Printing Hardware 157
6. "Black Box" Printing Systems 159

Chapter Nine
Using Large-Scale Book-On-Demand Services 164
1. Access to Large-Scale Hardware 164
2. Third-Party Book-On-Demand Business Categories 164
3. Submitting to Third-Party Book-on-Demand Services 166
4. More On Publishing Options 168
5. Conventional Publishers 169
6. Short-Run Printers and Digital Printers 169
7. Publishing Service Providers 171
8. Neo-Vanity Presses and Scammers 173
9. Contracts 175
10. Digital Warehouses 178

11. Specialty On-Demand Book Printing:
Hardcover Books and Picture Books 180
12. Full-color Books on Demand 181

Chapter Ten
Prototyping, Production and Business Decisions 185
1. Final Checks and Prototyping 185
2. Production Printing: Keeping Organized........................ 186
3. Taking Care of Business 188
4. Number Crunching 189
5. Fulfillment and Fulfillment Services 190
6. Fulfillment Software 190
7. Marketing 195

Backmatter
Appendix One *Book-On-Demand*
Printing As Appropriate Technology.................................. 198
Appendix Two *Sample Equipment Lists* 202
Appendix Three *Names and Numbers* 212
Index.. 224
Colophon.. 229
A Note At The End 230
About The Author 230

Preface to the Revised Edition

This new edition of *A Quick Guide To Book-On-Demand Printing* marks the first time in some years that I have given the whole guide a thorough housecleaning. I have deleted a lot of information that had become dated, or downright wrong, reorganized certain sections, and added coverage of new topics. Despite the deletions, the addition of new material has produced a net gain of about twenty percent in length, or about 12,000 new words of text. This edition of the guide is not quite as quick at the previous ones, but then it has further to go, through more information.

On the theory that my experience might be of some use to the reader, this preface seems a good spot to report on how I have used and am currently using book-on-demand, and how I'll be using it in the near future. When I first started out my small press business, the idea was to print all the books at home. For reasons that I think will become clear as you read the book, that proved impractical and I shifted over to mainly doing out-sourced on-demand printing.

As of this writing, my company, FoxAcre Press, has about 18 book in print, with more on the way. We use Lightning Source (discussed elsewhere in this book) to print nearly all of the books we sell. Even the copies of *this* book sold via Amazon.com and the other online booksellers are printed by Lightning Source. Copies of this guide purchased direct from FoxAcre I have, to date, made at home to prove to those who buy the book that it can be done.

As I write these words, my family is about to move to Germany for about three years. We'll be in an apartment without enough room for a big print operation, and it would obviously be impractical to mail books from there to U.S. customers. For these reasons, I will likely use Lightning Source to print nearly *all* copies of this book. In short, the *Quick Guide* will still be available, but I likely won't get to print and bind many copies of it myself.

There are limits to where and how you can do real production on-demand printing. However, you can print and bind *modest* number of books in the corner of one room, and I fully expect to do *some* printing and binding while I'm out of the country. It's too much fun to give up altogether. Besides, I am sure I will have some good reasons to make books—Advance Reading Copies of my next novel, or proof copies of new FoxAcre Press books, for example, or for various family projects. And I'm sure I'll come up with reasons to print copies of this guide as well.

Many readers will likely do what I do, and use a mix of at-home and third-party on-demand printing. It also became clear that many readers of this book will likewise find themselves in situations where at-home production printing at is not practical at all. They'll *have* to use an outside on-demand printer. This new edition therefore includes more extensive coverage of third-party on-demand printing.

Thus, the main addition to the text in this edition of the guide is in Chapter Nine, with a far more detailed discussion of doing out-sourced book-on-demand. If the rest of the book is about doing it all yourself, then Chapter Nine is about getting others to do it for you. I have also updated the *Names and Numbers* section, adding many new businesses—and deleting a number of others that are no longer with us. (I have also posted the current list at www.foxacre.com/bookpage/bod-address.htm, providing live links to most of the businesses listed. I will try to keep that online list updated, but that is no minor task.) I have also deleted descriptions of a number of pieces of equipment that are no longer available, and added reports on new hardware. Some parts of the book haven't changed much. There is very little new in the sections on hand-binding, or on page layout, for example. However, even these sections have received a going-over, and updates and corrections as needed.

As I prepared this new edition of the guide, it was a bit of a shock to realize how many of the businesses and services I reported on in the last edition are no longer with us. Part of this high mortality rate is just part of the normal life-cycle of business. Part of it is an echo of the whole dot-com boom and bust, when it seemed that anyone with an idea, good or bad, could get funding, and then all was swept away.

But part of it, I regret to say, is what might be called the true-believer syndrome. True believers fall in love with their business ideas, and don't much worry about what the customers might think. They pitch their overpriced e-book reading devices, their online services, their over-specialized tools and gadgets and gizmos, and never think to ask if anyone actually *wants* what they are offering. And they don't sweat the details, the fiddly day-to-day stuff of running their businesses, anywhere near as much as they should. (I don't have to look further than a mirror to spot someone guilty of that last charge.)

The death of so many true-believer run businesses in the past few years is a useful lesson for the would-be on-demand book publisher. You have to ask yourself if enough people will want the books you make to make the effort worthwhile. And it's not enough to ask—you also have to really do the research, and come up with an answer. The *good* news for book-on-demand is that the whole *idea* is making limited numbers of books. It could well be that "enough people" is fifty, or ten, or five—or one. One thing I truly believe in is book-on-demand printing. It's an exciting time to be working in the field. Read on, and then join in.

Roger MacBride Allen
Takoma Park, Maryland
August, 2002

PS—I could use help on typos, as this edition had to get out in a hurry, what with the trip to Germany and so forth. If you spot any mistakes in the text (and I'm sure they're there!), drop me a line at typo@foxacre.com.

Chapter One
Book on Demand: What, How, When, Why—And Who?

1. Introduction

The basic concept of printing books on demand is quite straightforward. The idea is to create finished printed material quickly, when and where it is needed. By printing only what is needed, and only when it is needed, you can save a tremendous amount of money, and revise quickly and efficiently.

On a practical level, book-on-demand printing consists of using a mix of new and old techniques to print books in small print runs as they are needed. Typically, the book pages are produced on a laser printer capable of double-sided, or duplex, printing. The covers are printed on a color ink-jet or laser printer. The pages are then bound into the covers, either by one of several hand techniques, or with any of a variety of binding machines, and then trimmed to size with a heavy-duty paper cutter.

The techniques used to print the pages and covers for book-on-demand books are new and evolving rapidly. On the other hand, the binding and cutting operations are essentially identical to those that have been used in print shops for the last hundred years or so, albeit in somewhat more automated form.

On-demand printing is more expensive than conventional printing when measured on a per-copy basis. However, on-demand printing makes it possible to print books only when they are needed. Updated and revised or even customized versions of a book can be produced quickly. Because print runs can be fine-tuned, and because additional print runs can be done with little or no make-ready time or cost, the economics of book-on-demand and conventional book printing are completely different. In situations involving short print runs, on-demand printing can be vastly more cost-effective.

Instead of having a warehouse full of a dated version of a book, a publisher can store an electronic version of the book in a computer, and keep it constantly updated, printing copies of the always-current information only when needed.

With book-on-demand, books need only be printed after they have already been paid for. The economics of this sort of "just-in-time" printing can compare favorably to those of "just-in-case" conventional offset printing, which often requires that the publisher print many more books than are needed. Storage and shipping cost money, and because this is so, "just in case" printed books soak up working capital, merely by existing before they are needed. As often as not, they are stored, shipped, shipped back as returns, (or never shipped at all) and then discarded. There are many costs aside from printing that must be borne "just in case" someone needs the books later. We'll explore this further in the next section of this chapter.

Because book-on-demand only prints books when they are needed, waste is cut back. It can thus wind up being cheaper and more profitable to print books using this more expensive processes.

As we shall see, there are large and small-scale versions of book-on-demand. The term "book-on-demand" can apply to a commercial shop using ultra-high-speed printers and automated binding equipment to bang out a hundred books an hour, or to someone in his or her basement who wants to run off ten copies of a book, and maybe is dreaming of something on a slightly larger scale. This guide will discuss both types of work, but is directed more at the home workshop. We'll also consider the various ways people in the smaller-scale end of the market can get access to the hardware used by the larger-scale operations.

2. The Economics of Book-On-Demand Printing

Let's take a look at how book-on-demand is different from and, for some jobs, better than conventional (usually offset) printing.

The core advantage is simple: book-on-demand allows a publisher to avoid all economies of scale. This sounds totally counterintuitive, but it really makes sense.

Conventional book printing is designed around the use of extremely expensive capital equipment (the printing presses), very expensive skilled labor (the printers) and very cheap raw materials (paper and ink bought at wholesale) to produce thousands of identical copies of the same book.

It is expensive to set up the equipment to print a book. It takes many of hours of skilled labor to design the layout, produce the plates used in offset printing, adjust the presses, and so on. That time and effort cost money.

To pull a number out of the air, let's say it costs $1,500 to set up to print 10,000 copies of a particular book. It might well cost that same $1,500 to set up for a 3,000 copy print run, or a 20,000 copy print run, or a 100,000 copy print run.

Once the presses are up and running, the system is highly automated, and a massive printing operation can be managed by a relative handful of people. Obviously, that drives labor costs down. If labor costs per unit manufactured falls, that leaves you with the cost of materials—and the wholesale cost of the paper, ink, and other materials that go into making a book are probably the smallest part of the expense of publishing most books. The cost per copy of additional copies is quite modest. Indeed, publishers often worry more about the cost of storing books than they do the cost of making books.

All of the above is reflected in the rate sheets of commercial printing plants. Typically, they charge a flat rate or hourly rate for setup, and then a per-copy rate that drops off sharply with the number of copies.

In short, it costs a lot to get set to print a book, but just a very small amount per unit to make more copies. Consider that book with the $1,500 setup cost. If the additional cost-per-book for materials were, say, $1.50, and we printed 1,000 copies, setup plus costs of the copies would be $1,500 plus $1,500, or $3,000, or $3.00 a book. If we printed 10,000, that would be $1,500 plus $15,000, or $16,500, or $1.65 a copy. That's a pretty tidy savings, and it ignores any volume discount the printer might offer.

These are made-up numbers, but they illustrate the sort of economies of scale offered in conventional publishing. Print enough copies of a book, and the books get pretty cheap to make.

But look at the numbers from the other direction. It's impossibly expensive to print just a few books. Go back to our costs of $1,500 setup and $1.50 per unit for materials, and think about printing five hundred copies. It costs $4.50 a copy. Printing fifty would cost $31.50 per copy!

Confronted with these sorts of numbers, and having the overhead of office rental, staff payroll, author royalties and so on, a conventional publisher won't even consider a book unless it has a chance to sell several thousand copies. A book that has something to say, but only to an audience of a few hundred people, can't possibly make sense to print.

Or consider the case of a book that would sell a hundred copies a month, year in and year out—a museum exhibit guide, for example. Such a book might be financially viable if one could print up a five or ten year's supply, so as to reach economies of scale, and then store them until they were needed. But the annual costs of storing that many books in a temperature-controlled, bonded, insured warehouse, complete with alert uniformed attendant, will quickly soak up any savings. (There are also tax reasons that make it wildly expensive to keep a lot of inventory on hand, but let's not get started on tax law.) Even if you could store the books for free, a book that's been in a shipping carton for ten years isn't going to look like new merchandise.

Or suppose the museum radically changed its exhibit two years after you printed a ten-year supply of the guide? Or suppose ten years of inflation made the price printed on the cover wholly impractical? Or a new donor made a huge donation and had the museum renamed to honor his mother? Or suppose any of a dozen unexpected things happened to make the guides dated? In such cases, the expense, trouble and risk involved with stockpiling a large number of books would more than offset the savings realized by printing a lot of copies.

Book-on-demand turns all this on its head. It uses moderately-priced capital equipment (computers, printers, and small-scale binding equipment), cheap or free labor (you) and reasonably priced raw materials (paper, toner, and adhesive purchased at retail or moderate discount) to produce books with a low setup cost and a moderate cost per unit (about $1.00 to $6.00 per book). Economies of scale are minor at best, but the initial setup cost can be as low as zero. Because books can be printed one at a time or ten at a time or a hundred at a time when and as needed, there is no need to maintain inventory. Updates and corrections to text can be made more or less instantly, between one copy of the book and the next.

With the typical conventionally printed book, once the original print run is sold, the book is no longer available. Books *can* be reprinted, but most never are. It isn't worth the time, effort, and expense. With book-on-demand printing, a book can stay in print forever. So long as the book is stored on disk, a fresh copy can be printed out whenever an order comes in, whether that happens once a month, once a year, or once a decade. Aside from the trivial expense of storing the computer file, the publisher can keep the book in "inventory" for something very close to free.

There is one other big factor to consider, however. Commercial publishers rarely own the presses that manufactures their books. They let someone else pay for the $10,000,000 printing plant. The publishers simply pay for the use of the presses.

Few book-on-demand publishers will be able to avoid spending substantial money on hardware. Unless you set up an absolute bare-bones operation that cuts out practically everything but glue, paintbrushes, clamps, and courage, (in which case you might spend $100 or less) your initial capital investment could be substantial. It might be several hundred dollars, or even a few thousand. Your setup cost per title can be small, and your operating costs should be pretty low. But the first step can be a doozy.

3. Purposes of This Guide

This is *not* intended as an all-inclusive how-to-do-it guide. It is instead intended to give a basic overview of what's involved in book-on-demand. It won't give you all the answers, but it should equip you to ask the right questions. It's more like a *what*-to-do guide, discussing what you need to understand and the tasks you'll need to accomplish if you want to make books. You might not be an expert after reading this book, but you'll be able to make a good start on book-on-demand printing.

Creating books is a complicated business, but much of what it involves will be fairly common knowledge to most people considering such a project. (For example, the reader is assumed to know what a word processing program is.) This book focuses on the things I found hardest to learn about, the topics where I really had to dig to find the information. There are lots of topics this book mentions only in passing, or not at all, because they are discussed in detail in other, easy-to-find, books or other places.

We'll look at the business side of book-on-demand publishing in several different contexts. Setting up your printing operation—whether for business, charity or pleasure—will require a lot of hard-edged decisions about money. We'll focus on the specific issues germane to printing and publishing, as the more general issues are discussed in lots of books about starting your own business. This is a book about how to make books.

This book cannot provide an exhaustive report on every possible variation on book-on-demand printing. However, it will offer lots of leads on other sources of information for just about every topic it mentions. (See the *Names and Numbers* section in the back of the book. People, companies, products, and publications mentioned in *italic* on first reference in a section of this guide are listed there with contact information.) As I wrote the book, I found that locating *sources* of information was 99 percent of the research battle. That accomplished, getting the information itself was duck soup.

This guide will help you decide if you're up to doing some book-on-demand work in your basement, in terms of finances, patience, mechanical ability, and so on. It will point you in the right direction for much of the research you need to do, and get you thinking about what hardware you do and don't need. It will discuss various other means of getting published. Maybe book-on-demand isn't right for you. Therefore we'll discuss the alternatives that might suit you better.

I have deliberately left in certain somewhat redundant passages in this book, on the assumption that many readers will flip through and find the sections that interest them, but never read the book cover to cover, or else will read the whole book through once, but then, at some later date, check back on the items they found of interest. In order to accommodate such readers, if a point is important in more than one context, it is mentioned in more than one context.

I update this book frequently, as I learn more and gain more experience in the field. The revision date for this book is listed on the publishing information page at the front of the book.

There are lots of other ways to produce the pages of a book besides the ones we'll discuss here. But for the vast majority of home book-on-demand printers, the best way to print pages, by a wide margin, is with a fast laser printer. This book assumes that is what the reader is using, although much of what is discussed, for example regarding paper-cutting and binding, will apply no matter what printing techniques (of those available to the home book-on-demand printer) are used. See Rupert Evans' *Book-On-Demand Publishing* for a detailed discussion of other possibilities for printing pages.

4. How This Guide Got Here and Where It's Going

I write science fiction novels for a living. One of my best books received exactly one U.S. edition, and was never seen again on these shores. I couldn't get it reprinted to save my life, so I decided to do the job myself. I started FoxAcre Press so as to give my older work, and that of other writers, a chance to find new readers. I started taking notes on the subject of on-demand printing in the process of gathering information for that business. After a while, I realized that my notes were getting rather extensive. Those notes evolved into the first edition of this guide.

Writing this guide helped me put my own thoughts about

my business plan in order. It focused me on some of the decisions I needed to make, and experiments I needed to try out. In a sense, the first version of this book was a shopping list of the things I had to track down and figure out, including quite a number of issues I didn't think to consider when I first started out.

It was and is also something like a textbook, written by the student as he learns. That learning is still going on, and this edition contains a good deal of new information.

One irony of my research is that it happened just at the time when radical changes were taking place in the world of book printing and publishing. Large-scale book-on-demand printing was coming into its own, and Amazon.com and the other online booksellers were making sales of small-press books vastly easier.

I did enough research, and crunched enough numbers, to realize that, given my particular situation, it didn't make a great deal of sense for me to follow my original business plan and rely on at-home on-demand printing for all the books I wanted to publish. It would make more sense to use one of the on-demand printing services, and let *them* do the printing. By the time I was done researching, I was using printing and distribution systems that literally did not exist when I started.

The same is true regarding the technology used to design and print the books. Back when I started researching, all the products—the computers, the printers, the software—were nearly good enough, nearly fast enough, nearly cheap enough. But prices on computers and printers have dropped while their abilities have vastly increased. All sorts of things that were just barely possible, or just barely affordable, are easy to do, and much more inexpensive. Unfortunately, the prices on page-layout software, and on binding equipment, have not gone down, but there's enough good news on prices and capability that it almost doesn't matter.

In similar vein, my primary research tool, the Internet, barely existed when I started. There are now so many websites, tools, resources and so on available that the Internet must be considered an essential took for anyone doing book-on-demand printing.

5. Sources for Information In This Guide

While the information in this guide came from many sources, the single best and most useful source was a book by Rupert Evans. The book is *Book-On-Demand Publishing*, from Flash Books. If you are interested in book-on-demand printing, you should get this book. While its information on computer software is dated, the chapters on paper and binding are chock-full of useful ideas and information that's not going to get old any time soon. I would have saved myself a great deal of time, frustration, and money if I had found this book earlier on in my book-on-demand career. Buy yourself a copy.

A secondary source of information on book-on-demand was *Don Lancaster*, and his company, *Synergetics*. Mr. Lancaster's web site, www.tinaja.com, is quite helpful and well-organized. Many of Mr. Lancaster's articles are available as free downloads in the Acrobat PDF format. The book-on-demand material from Mr. Lancaster consists in large part of reprints of various of his magazine articles. A given column might cover three or four topics, only one or two of which might be germane to book-on-demand. Much of the material is recycled, nearly word for word, several times, from one column to the next, and, as of this writing, a lot of it is years out of date. Much of the hardware Mr. Lancaster discusses is no longer available. Nonetheless I did find a great deal of value in his columns.

There is one omission in both Dr. Evans and Mr. Lancaster's discussions of book-on-demand: any detailed examination of small commercial-grade book-binding equipment.

While such machinery is beyond the needs and means of most book-on-demand shops, it is useful to understand its operation, and the meaning of the terms used. After experimenting, rather successfully, with variants on most of the hand-binding techniques I discuss in this book, I tracked down a good table-top binding machine going cheap. I decided that the improved quality and consistency of the binds produced would be worth the expense involved, but it took close to a year of digging before I really understood binding machines well enough to make that decision. The machines themselves are simple enough. What took all the digging was finding the manufacturers, getting the spec sheets, and finding places where I could get a look at the machines themselves, rather than pretty pictures in a brochure.

There is a promising trend toward relatively (and I stress the word relatively) affordable book-binding equipment. While these machines are still well beyond the budget of a home user, they are certainly well within reach of a church, a community center, or a school.

I have therefore included a fairly extensive discussion of these machines, both in respect to their general mode of operation, and in respect to the features and flaws of various models. This information is based on material provided by the manufacturers, and by a limited amount of hands-on inspection of the hardware, the many product brochures I managed to get from various companies (though it took several tries to get some of them), articles in magazines that mentioned a product that might be of use, searches all over the Internet, visits to trade shows, and, last but not least, my own book-on-demand printing experience, which is not tremendous, but not inconsiderable. The good news is that finding out things has gotten much easier. It used to be that hardly any of the companies linked to book-on-demand had any sort of website, and the websites that did exist were far from useful. Nowadays, nearly every machine, product, company or service mentioned in this guide has a website. Some of them are fluff and pretty pictures, but a lot of them provide solid and clear information.

Another source for the first editions of this book was *Flash* Magazine. While this new edition still includes information from that source, much of it has been superceded by advances in the field since *Flash* ceased publication. *Flash* was a real voice in the book-on-demand world. Their main focus is—or was—on doing more with less: certainly a subject dear to the hearts of anyone doing book-on-demand. Much of what they dealt with pertained directly to book-on-demand. *Flash* was a intriguing source of information, but not particularly reliable when it comes to being available. It took some real doing, and far too long, to get them to ship me my copy of Rupert Evans' invaluable book.

Their business was quite literally struck by lightning not long after they converted from a conventional printed magazine to an on-line publication, and the lightning wrecked much of their equipment. In their first three or four years of "monthly" web publication, they have managed to get out two issues. As this edition of the *Quick Guide* goes to press, their website is once again promising a re-launch of the magazine. This is far from the first time this has been promised, but I hope that manage it this time, and they get back on their feet and publish again soon.

Even if no new issues of *Flash* are forthcoming, their back issues (which are available for sale, on paper or on CD-ROMs, through the web site or via phone or mail) are full of intriguing articles. The product reviews aren't going to be much help, but there is a lot of discussion of tips, tricks, and techniques that should hold up pretty well.

Getting back to sources, other useful fonts of information included *Independent Publisher* Magazine (also now fully online with no print edition) and *Adobe* Magazine. (In both cases, the ads are often at least as useful as the editorial content.) I also got much use from Dan Poynter's book, *The Self-Publishing Manual*, published by *Para Publishing*.

Regarding my own hands-on experience, I have done one form or another of book-on-demand printing, with limited to good success, for about twenty or thirty titles. I've written books, done book layouts, designed and printed covers, and cut and bound and trimmed books.

I've flinched as I realized the book was crooked in the cutter a half-second too late, moaned as the binder melted my spine lettering, rolled my eyes when I discovered that I had the cover on upside-down and the pages in the wrong order, and sworn a blue streak when the pages fell out of my "perfect" bound book.

If you learn from your mistakes, then there can be no question but that I've had a lot of chances to learn.

Now it's your turn.

Chapter Two
Publishing Realities and Publishing Choices

Before we get down to working with paper, glue, toner and ink, let's consider the arguments for and against book-on-demand. As it will be easier to judge if book-on-demand is right for you if you know what your other options are like, we'll take a look at the many ways of getting someone else to do the work of making books.

In this chapter, we will look at a number of different publishing alternatives in order to place book-on-demand in some sort of context, and to see what sort of other choices are out there.

I will explore these other options at some length in this chapter, because it is important for anyone considering anything beyond the smallest-scale book-on-demand operation to understand all the other ways there are to make books and get them to readers. Later in the book, once we've explored the work of making books on demand and have a better grasp on the mechanics of the process we'll come back and take another look at the business side of things.

If you want to dive right into the how-to part of the book, skip to Chapter Three.

1. Who Should Consider Book-On-Demand Printing

One answer would be: people who have worthwhile books to print, but who don't have (and/or perhaps don't want) a commercial publisher for the book. Another answer would be: people with books that have small but valid audiences, or books they expect to sell at a slow but steady rate over time, and/or books that might require frequent updates and revisions.

Any would-be book-on-demand printer needs to consider how he or she will sell or market or give away his or her books.

Who is going to want your books? How do you get the books to them? How much, if anything, will you charge for them? Can you accept the financial risk, and the emotional let-down, if things don't work out? If you can deal with such questions realistically, and come up with reasonable, practical answers to them, then printing your own books might make sense.

2. Who SHOULDN'T Consider Book-On-Demand Printing

Who should stay away? For starters, all those whose main motive is the need to prove something to the publishing powers-that-be who rejected their work.

Being rejected by mainstream publishers is not, in and of itself, an argument against printing your own work. If your book has been turned down by every publisher in New York, but you want to print copies so people can read your work, that's one thing. Print the copies, sell and/or distribute them, and enjoy yourself.

But if you want to print copies in order to show up all those smart-aleck editors and publishers, if you're fantasizing about making the big publishing houses come crawling to you after your self-published book has sold a million copies, forget it. Either scale back your expectations, or consider another hobby.

In the old days, the only option available to an egomaniac with an unpublishable manuscript was a vanity press, as discussed a bit further down in the chapter. Nowadays, there are a whole range of options for producing books that no one wants to read—including book-on-demand printing.

Book-on-demand might be a novel way to manufacture books, but it doesn't solve your distribution problems, and doesn't make anyone want to read your words. Unless you do your homework very carefully, and market your book very well, and push very hard, you're not going to sell any more books via book-on-demand than via the vanity press route—and vanity press basically doesn't sell any books at all. If you're not careful, instead of a bunch of expensive and unsalable books cluttering up your basement, you'll have a bunch of expensive and unused printing equipment cluttering it up.

3. Is Doing Book-On-Demand Worthwhile?

Gathering the equipment, learning the procedures, and simply doing the job of printing and binding your own books is a lot of work, and will cost you money, maybe a lot of money. What would make such an expenditure of time and money worthwhile *for you*?

Is untold wealth the only thing that's going to do it? Maybe making a buck your only motive for doing book-on-demand printing. Or perhaps money is not any part of what motivates you. You might want to print the books and give them away to friends, members of your church, relations, enemies, random strangers, co-workers, and/or tollbooth attendants. Maybe you will derive satisfaction from spreading your words around, rather than from making a profit. Or maybe you just want the satisfaction of putting a book into print, the pleasure of making something yourself.

Most people will fall somewhere between these two extremes of doing it strictly for the money and strictly for pleasure. That's about where I am in regard to this book. I don't expect to sell thousands of copies. But I would like to have this guide, and the other FoxAcre titles, available to readers who are interested. If I can turn a dollar on the project, or at least break even, fine. But if making money were my primary goal, I'll try and come up with something a bit more practical as an investment. I regard this book as a modest success because *some* people have purchased it, even if millions of people haven't.

And besides, I can use my book-on-demand skills and hardware for other purposes. For example, many publishers do Advance Reading Copies (ARCs) of their books. These are circulated to reviewers, and to overseas publishers who might be interested in doing foreign editions. But publishers don't do ARCs of every book, they don't get ARCs to everyone the writer would want, and they often get them out too late to do much good. I decided to solve those problems.

Without going into too many details, there was one book of mine where I went all out getting my own ARCs done. I hired a printer and paid top dollar and rush fees and got a very glitzy cover done up. This was back in the days before the costs of on-demand printing dropped dramatically. From start to finish, it cost me something like $1,800 to get fifty copies done. (Today I could probably do the whole deal, including the cover art, for under $600.) I simply gave those fifty copies away—prior to publication of the commercial edition. That sounds like a money-losing proposition, but I gave the ARCs to foreign publishers, to reviewers, to editors, and so forth. The early visibility did me a lot of good, and I made my money back many times over, mostly because the advance reading copies helped sell foreign rights to the book.

There are lots of ways books can be successfully published, and yet not make money. There are plenty of good reasons for doing books that don't involve trying to sell them at a profit. You might be considering a parts catalog, or an employee handbook, or a book of family remembrances, or a cookbook your church wants to put out. Such a project might never make a dime, and yet be a great success. Or you might know for certain that you have *the* book for a certain market, and that you can sell as many books as you make for forty bucks a throw.

Whatever your goal, do your homework. Study the market. Define what you would regard as success. Figure out how bad failure could be, and ask yourself honestly if your wallet (and your ego) could handle such a failure.

Depending on your situation, spending the time and effort to get set to do book-on-demand might or might not be worth it for one small batch of books, one super-short run of a single title. If you expect to do several titles over time, or to print more books, or repeated small batches of books, it becomes a lot easier to justify the investment. Figure out how many titles you expect to publish, and how often you want to reprint how many copies.

Nor is there any reason to assume that a book-on-demand printer has to be a private individual. Indeed, a book-on-demand operation might make more sense connected to a school or a library, so it was available to community members who took a training course. (And have those community members sign release forms before they get near the 350-degree hot glue and the two-foot-long razor-sharp paper-cutting blades!) A writer's center, a community center, a retirement home full of people with memoirs to write, might all be worthwhile venues.

Once you're set up, printing and binding your own books is fun, satisfying, and not all that hard. But it requires a substantial investment of time and money to do it properly. Consider carefully before you go for it.

4. Publishers, Conventional and Otherwise

The odds are pretty good that most readers of this book have one or more books they want to get published. The odds are likewise pretty fair that they have tried to get their works in print, but haven't had much luck.

The sad and unpleasant truth is that most books that are unpublished, *deserve* to be unpublished. They just aren't good enough to bother going to the time and effort of commercial publication. Some unpublished books are not bad—but they just aren't good enough. They are simply too dull, or too derivative, or too predictable. At the other end of the spectrum, some unpublished (and published) books are plainly the products of unhinged minds.

Based on my own personal anecdotal research into the question, consisting mostly of lots of conversations with lots of editors, my best estimate is that only one out of a thousand unsolicited manuscripts gets published. That ratio pertains to both novels and short fiction, and I'd expect the statistics for poetry and nonfiction would be broadly comparable.

I've often asked editors how many more books they would publish if the only criterion were quality; if they didn't have to worry about sales figures, print runs, profit margins, and so on. The usual answer I get is that they wouldn't publish all that many more titles, simply because very few of the unpublished books are good enough to make it worthwhile. Some say they wouldn't publish any more at all. Others allow as how they might print maybe fifteen percent more material.

This is not to say that all books that get passed over, deserve to be passed over—or that only deserving books see print. We've all read God-awful books and wondered what on Earth possessed anyone to publish such rubbish. We've all muttered that we could do better, and that is quite possibly the case.

I write books for a living, and I go back and forth between thinking that any damn fool can write a book, and thinking that writing is an intricate and delicate skill that few are capable of, and that even fewer master. Suffice to say that most of the unpublished manuscripts I've read support the latter, rather than the former, idea.

Wherever the truth lies on that question, I do know this: *selling* books is a totally different skill than writing them, and, frequently, a tougher one to learn. Even if you print five hundred copies, or five hundred thousand copies, of a book that deserves the Pulitzer, that doesn't mean anything, all by itself. You have to market, sell, store, and ship those copies, and get them in the hands of readers. That is what commercial publishers know how to do, at least some of the time. It is what other kinds of publishers *claim* to know how to do.

What follows is a brief discussion of the various forms of publication, as seen from the writer's point of view. Here, we're concerned with the basics of the business structure for each form of publication, not with editorial submission policy or contract negotiation or any of that. There are endless books, articles, and so on about how to get published that cover all such matters in great depth.

Here we're just taking a quick look at the business models for each type of publication, identifying who is responsible for what in conventional, vanity, private, and self-publishing, and the various variants therein thereof.

5. Conventional Publication

I will devote much of this section to what conventional publishers do *not* do, as such issues will likely have some significance for many authors considering book-on-demand. A writer who has failed to make a deal with a big-time publisher and is looking for alternatives might well come across certain companies and individuals who claim to be part of the commercial publishing world, but aren't. Watch out for them.

I often teach writing classes, and when the subject rolls around to the business side of commercial publishing, one of the first rules for writers and money I lay out is this: *the money should always move toward the writer*. Any business arrangement, presenting itself as part of the system of conventional publication, but not following this rule, is all but certainly a scam of one sort or another. Of the remaining small fraction of cases that aren't scams, many or most are likely to be deals offered by rank amateurs who don't know the business themselves.

When I teach writing, I make it a rule to be paid by the institution that hires me, and *never* directly by the student. I established this rule so that I would never have a direct financial relationship with the person whose work I was evaluating. How could the student trust my objectivity if they were paying for my opinion? I would have a vested interest in giving the advice that was likely to get them to pay me more.

For similar reasons, I have for years advised my writing students and other would-be writers to steer well clear of any so-called agent, freelance editor or publisher who offers to read over a manuscript, or edit it, or publish it, for a fee. I would no more do business with such people than I would trust a prospective employer who wanted me to pay him for reading my résumé.

Legitimate agents, publishers, editors and so on *never* charge writers for their services. They make their livings off the proceeds of books that sell to a paying public.

Any organization representing itself as a commercial publisher that requests any sort of fee from the author should be avoided. Anyone claiming to be a free-lance editor who offers to read a manuscript for a fee should be avoided. Anyone claiming to be an agent who requests a fee should be avoided.

Such individuals and organizations almost invariably make all, or the vast majority, of their money from such fees, and not from the proceeds derived from the commercial publication of books. Some merely skirt the edge of fraud, while others tumble right over into it. Fee-charging commercial publishers (or, more accurately, vanity presses), fee-charging agents, and fee-charging editors make their living, not by helping writers succeed, but by profiting off writers' continuing failure. The exceptions to this rule are so rare as to be the stuff of legend.

Legitimate commercial publishers, agents, and editors will, quite sensibly, have nothing to do with these scam artists. Contact with the scammers can do a writer no good. For example, an author represented by a fee-paying agent is not improving her odds of seeing print. Because many real publishers will not even return such an agent's calls, such representation probably *reduces* the odds of publication.

All that being made clear, what is it, exactly, that commercial publishers do, aside from not charging for their services?

For starters, they are the bankers of the business. They take most of the financial risk and reap most of the financial reward. They put up money for the writer's advance, for the printing bill, for the editor's paycheck, for shipping and storage of the books. Through all sorts of complicated arrangements, they in effect extend credit to book stores. (At any given time, the publishers are likely not to have yet received payment for up to forty percent of the books in your neighborhood bookshop. The bookstore might not pay the publisher for a particular book until months after a customer has paid the bookstore for it.) In exchange for taking on all this risk and exposure, publishers make sure to put themselves first in line for any profits a book might earn.

Legitimate conventional publishers sell wholesale to distributors or bookstores, keep most of the money for themselves (and use it to pay lots of bills), and pay royalties to authors, based, in one way or another, on either the number of copies sold, or the number of copies they expected to sell. Typical royalty rates are about six to eight percent for paperbacks, and ten percent for hardcover books.

How do the other professionals in the field get paid, if authors don't pay them fees?

Legitimate agents negotiate deals for their authors, manage various business issues, and take a percentage (usually ten or fifteen percent), of payments received from the publisher. Agents sometimes pass through charges for mailing, photocopying and so on, but such charges aren't too popular with authors.

Legitimate editors are paid by the publisher, either on a full-time or free-lance basis. Editorial payroll, and all the other expenses of the publisher, are paid for with the income derived by selling books (and subsidiary publishing rights, and other forms of intellectual property).

Many, if not most, commercially published books fail to turn a profit, but the money-making titles make up for the money-losers. Publishers are willing to take the risk of a money-losing title for several reason. First, even if only one out of three books (to invent a statistic) turns a profit, no one can be sure which of a given three it will be. The publishing house is gambling—but gambling with its own money, and in a game it understands perfectly well.

A publisher might take a loss on a first book in hopes of developing an audience, or improving the author's skill, or for other, more Byzantine, reasons. And, in the grand scheme of things, not that much money is at stake with a given title. To invent more numbers, if Bigtime Books publishes ten books, and eight of the titles lose five thousand dollars each, but the ninth earns fifty thousand and the tenth makes hundreds of thousands of dollars, or even millions, the loss of forty grand generated by eight losers will look like an acceptable cost of doing business.

When a legitimate publisher contracts with an author, either directly or through an agent, the contract is structured in such a way that, if the publisher accepts the manuscript and publishes it, the publisher, and not the writer, is liable for any losses that may accrue. In other words, if yours was one of the eight money-losers for Bigtime Books, Bigtime can't and won't ask for the money back. Leaving aside complicated and rare contractual disputes, once an author is paid, the author does not have to pay any refunds.

Aside from financing the process of making and selling books, commercial publishers serve as the middlemen, the prime contractors who make things happen. They don't create books (the writers do), they don't print books (they get printing companies to do it for them), they don't even store the books (they have contracts with warehousing companies).

What they do themselves, for the most part, are four things (and, granted, they often don't do them well). The basic tasks of publishing are these:

- The selection and acceptance of books from those submitted, followed by the entire process of editing, copy-editing and proofreading.
- The design and layout of the book's cover and interior, setting type, and procuring art and design for the cover and any interior art. (This set of tasks is also often partially or wholly sub-contracted out.)
- The management of legal and business matters, such as getting enforceable contracts signed, registering copyrights and ISBNs, and tracking sales and accounts.
- The complex process of getting books into the hands of the reading public.

The last of these four tasks, getting books to the public, seems as if it should be a straightforward job, but it isn't. Problems with drop-shipment, muddles over the reserves against returns, jumps in warehousing expenses, crises in the distribution channel, jiggering of discounts, and a dozen other headaches crop up every day. It's a tough, complicated, on-going process, and it's very much a full-time task for professionals.

Commercial publication is the best way to get a book out

there. However, from the writer's point of view, it is far from a bed of roses. For starters, commercial publishers don't always know what they are doing. Publishers might be professionals, but they aren't always competent.

One of my books—no, make it two—were not so much published by the publishers as sabotaged. So few copies reached the reading public, and those were so badly mis-marketed, that the number that did sell must have done so as a result of clerical error. But the publisher did this not through malevolence, but through idiocy, incompetence, indifference, the backwash of office politics, and through the unwillingness of underlings to tell the boss he or she was wrong.

Nor are publishers always all that well-organized. I have, as of this date, published eighteen novels. With one exception, for every single title of those eighteen, there has been at least one error in payment to me. (Book eighteen is bound to join the club any time now.) In every single case, the errors have been in the publisher's favor.

With one exception, I have found all the people and organizations I have dealt with in publishing to be honest and honorable in their dealings. Sooner or later, all of the payment muddles (with that one same exception) were straightened out by the publishers—after the money in question had been drawing interests in their accounts instead of mine for an extra two or three or six months. But when all the errors are to the other side's advantage, it does suggest a certain institutional bias toward sloppy management and creative accounting. Foul-ups reward the publisher, and thus making sure they don't happen again isn't a big priority.

Besides, unless your name is Grisham or King or Hemingway or Steele, you need the publisher more than the publisher needs you. The laws of supply and demand apply in publishing. There is an oversupply of writers, and there is excessive demand for publishing slots. There are lots of would-be writers who would work for nothing, or close to it, in order to see their words in print. This, of course, drives down the rates paid to all the other writers, because both sides know the publisher can always find someone willing to work cheaper. Result: there is a certain amount of game-playing in the world of publishing, and the average author unquestionably has fewer chips on the table than the publisher.

Even when everything goes right, success is far from guaranteed. The most likely fate for a published book by a new author is to be published in one edition, to receive few or no reviews, to sell a few thousand or tens of thousands of copies, and then to vanish without a trace. Publishing can be a lot like throwing a very small pebble into a very large ocean of books.

However, despite the foregoing gloom and doom, it is important to emphasize that, for the vast majority of books, conventional commercial publication is nonetheless the first and best choice. If it is an available option, take it. It is best for the book, and best for the author. If you have the choice between Random House or book-on-demand, go with Random House. The smallest conventional publishing house, screwing up big time, is probably going to sell more copies of your book than you can yourself, even if you do everything right. This is not universally true, but it is darn close.

But for books wherein commercial publication is not possible, there are other options. The previous discussion might imply there is only one good way, one right way, to get published, but the world is not that clear-cut, that black and white. There are forms of publishing that might be likened to shades of gray, or even splashes of color. These options are not wholly distinct from each other, but merge, one into the other, across a spectrum of possibility. Let's look them over—starting with the one that is the darkest shade of grey, verging on absolute black.

6. Vanity Press

As might be gathered from the preceding discussion, just writing and printing books can be the easiest part of publishing. Selling the books, getting them into the hands of readers, is another, and often greater, challenge. Unfortunately, there is a whole industry based on making people *think* that anyone can do the marketing, packaging, shipping, selling and inventory-management that commercial publishers do. It is the sleazy and unpleasant business that likes to call itself "subsidy publishing," but should be known by the name "vanity press." It is an industry based on deluding people.

How does it work? Simple. Give a vanity press a manuscript, any manuscript, no matter how bad, and a large enough check, and the press will promise to "publish" your book.

As we have seen, a normal publisher *pays you* for publishing your work, and takes on the job of designing, printing, promoting, marketing, shipping and selling your book. A vanity press publisher *charges you* for publishing your book, and for whatever else it can get away with, and "publishes" it by assuring you they will work hand-in-hand with you to promote and sell the work. In the vast majority of cases they don't actually do any promotion or marketing at all.

In a typical deal, a vanity press will agree with a writer to print a thousand copies of his book, with half the print run going to the author, and half to be "marketed" by the brilliant professionals in the vanity publisher's sales department. The sales department might not even exist, but if it does, rest assured it doesn't sell many books.

They ship five hundred copies of your book to your house, where they will clutter up the basement, and, as soon as they can get away with it, they pulp or dump their five hundred copies, and sadly report to you, the author, that sales were slow. The vanity presses skate right up to edge of fraud. Their primary source of income is not book sales, but fees paid to them by authors. In short, vanity presses feed on the weakness—and vanity—of their customer/victims.

The good news is that the old-line vanity press has been weakened, perhaps fatally, by the rise of the legit—and not-so legit—large-scale book-on-demand printers, which we'll discuss in Chapter Nine. The bad news is that I have no doubt that the vanity presses will learn to adapt to new circumstances, and shaft whole new generations of writers.

7. Private Publishing

This is a term I have invented, but I think it's a useful one. A private publisher in essence does what a vanity press does, except a private publisher does it honestly. That is no small exception. Private publishers offer virtually the same service as vanity presses, but don't *pretend* to be doing something else. Vanity presses make their living by conning writers, and trade on vanity and false hope. Private presses provide a service, and charge for it. One other point about private publishers: they are rare.

To define private publishing another way, it is third-party self-publishing, wherein you, the writer, in effect hire someone to self-publish your book for you. The writer delivers a manuscript, and pays for services rendered. The private publisher may offer editorial services, or may simply take the manuscript as delivered, and then, for a fee, do the design and layout, contract for printing, register the book's ISBN, and produce finished books ready for sale.

Maybe the publisher ships those books to the writer's garage or basement, making no bones about the fact that it is up to the writer to sell them. Maybe the private publisher does get involved in distribution and promotion—for a fee, or a percentage.

8. Conventional Self-Publishing

The only real difference between a private citizen and a publishing house is that a publisher has purchased some International Standard Book Numbers (ISBNs). However, as discussed elsewhere in this book, a private citizen can get a starter set of ISBNs for something like $225. I did. So now I'm a publisher.

As noted above, what a publisher does, basically, is put up the money, and serve as the prime contractor who sub-contracts out the work to others. As the name implies, in self-publishing, you become your own publisher, your own prime contractor. You do all things a publisher does, or else you sub-contract for typesetting, design, printing distribution, and so on.

In this day of desktop publishing, a typical self-publisher will do his or her own design and layout, and send the completed interior and exterior layout on disk or via the Internet to a printer specializing in short-run printing. (For some reason, lots of these specialty printer are in Michigan.) The printer will ship the books to the self-publisher, who will then be responsible for promoting, distributing, and selling the book.

The huge and obvious disadvantage to this process is that the author is left with the task of moving books out the door, a task that not even commercial publishers do all that well. Few self-published authors get far going bookstore to bookstore, asking the clerk behind the counter if the manager would be interested in a few copies of a self-published book, though it can be done—occasionally.

Finding distributors, negotiating returns, getting bookstores to pay up, and doing the actual packing and shipping can suck up more time than it takes to write the book. Dan Poynter of *Para Publishing* wrote *The Self-Publishing Manual* and also offers a great deal of information on this subject through his web site.

A big potential advantage of self-publishing is that all the proceeds from every book can go right into the self-publisher's pocket. The writer of a conventionally published book with a twenty-dollar price tag and a ten percent royalty rate earns two bucks a book. If a self-published book has a twenty-dollar cover price, the self-publisher might sell copies to bookstores for a wholesale cost of from ten to twelve dollars and get to keep all of it. If she sells that same book by mail order direct to the reader for the cover price, she gets to keep all twenty. Some books will do better sold at wholesale, while others will do better with direct sales. Some self-published authors do very well. Others have their problems.

Obviously, a conventionally published book (with the author getting a payment of ten percent of the cover price of each book) will have to sell ten times as many copies to provide as good a gross return as a self-published book where the author takes one hundred percent. However, the downside of self-publishing is that most conventionally published books can do just that with no trouble at all. And the author of a conventionally published book doesn't have to take out a bank loan to pay the printer's bill.

9. Third-Party Book-On-Demand Publishing Services

There are a number of business types that fall under this heading, and we'll examine them in greater detail in Chapter Nine. For the purposes of the present discussion, suffice to say there are businesses that use book-on-demand technology, similar to but larger-scale than the techniques we'll learn in this book. These companies will take on part or all of the tasks required of a publisher, and charge you for the service. Their businesses are structured in a number of ways. Some are entirely legitimate and upstanding businesses, some are in a bit of a grey area, and some are downright slimy.

Suffice to say that there are business that will, for a fee, do any or all of the following: edit your book, design your covers, print your book, store it electronically, give it an ISBN and barcode, store it, ship it, promote it (or pretend to promote it) put on a web page, put it on a web page someone might actually see, and, just maybe, send you money when—or if—someone buys a copy. There are lots of ways to do business. We'll discuss them in more detail after we've learned more about the mechanics of making on-demand books ourselves.

10. Web-Based Electronic Books

The web-based electronic book is the publishing variant that requires the least effort above and beyond the actual writing of the book itself. It is, in essence, nothing more than the manuscript itself, posted to a web site or stored in some sort of electronic form. A typical electronic book, or e-book, is offered as something that can either be read off the screen of an Internet web site, or downloaded from the site, either for free or for some sort of fee. Maybe a free sample is available, the first chapter or so, with the rest of the book available only after payment in full.

The "publisher" of such a book does not have to deal with paper, ink, binding machines, or any other aspect of manufacture, because nothing is manufactured. If the software allows it, the book can be printed out, but this is often (deliberately) made to be an awkward and expensive process. But if I download a book for "free" from the Internet, and then print it out, it might well take up a big hunk of connect time, and then tie up my computer and printer for an hour or more. The cost of the paper and toner for a three-hundred page book will cost me more than a real paperback book, and the printed e-book won't have covers or a binding. One could of course read the book on the screen, but there are only about a zillion studies that show how much people dislike doing that, and how much harder it is than reading a book.

Furthermore, it can be tough to control sales of an e-book. If I pay ten dollars to download it from your site, and then start handing out pirated copies of that file to all my friends, you won't see anything beyond that ten bucks, even if I upload copies of your book to a dozen download sites. There are security techniques that protect against this, but they all make the process just that little bit more cumbersome.

Bibliobytes, formerly at www.bb.com, was a web e-book service that used advertising to support free on-line viewing of books. You read the book, page by page, on the screen, while the ads appeared elsewhere on the same screen. Authors received a cut of the ad revenue. However, Bibliobytes ran into a lot of bad luck and folded. I suspect part of what did them in was not the virus that infected their web servers, or the dot-com crash, or the business slowdown in fall 2001. They were also hurt because no one quite understood their business model—and because, in my opinion, fairly or not, the act of giving away their product (the texts of books) made it look like their texts weren't *worth* anything.

There is, after all, the psychological question of perceived value. If a product or service is offered for a price, the buyer will likely assume that is the price it is worth. If it is offered for free, for nothing, the "buyer" assumes it is *worth* nothing. And, with no disrespect intended to Bibliobytes, by this time of day, it doesn't take much consumer savvy to realize just how close to true that perception is when it comes to books on the Internet. Anyone cruising the net will quickly realize that most books available for free on the World Wide Web are books that no publisher was willing to pick up. In cold hard fact, what you have with most e-books are rejects and failures, offered for what they are worth in the market: nothing.

There are exceptions to this. But even when this perception is unfair and incorrect, it is still out there. Something similar applies to magazines that move onto the Web. When I heard *Flash* was going up on the web, I knew enough about why previous magazines had done the same to know that it was not good news. Despite all the happy press-release talk to the contrary, I knew that no one moves a *successful* book or a magazine from a print version to the web, with no print version available.

Even if you charge for your services, the sheer number of freebies and near-freebies out there on the web drives down prices. If forty-seven similar objects are offered for free or close to it, the average citizen is going to expect the forty-eighth example to be free as well, and will be most unwilling to pay if it isn't.

Interestingly enough, and turning this perceived-value problem on its head, I have heard of at least two cases where offering unlimited "free sample" downloading of e-book type material from a web site has produced sharp increases in the sale of real books through the web site. People like the words they read, but do the math and realize that buying a real book for cash money is a better deal than a full download of the text for "free." I doubt that technique would work for all forms of books. (In fact I know it won't—one of the businesses using this technique is either already dead or so completely on life support it's impossible to say if it could still breathe on its own.) But in some cases—basically academic books—this idea works. Or at least it *seems* to work. Check back in a couple of years.

11. Proprietary-Format E-Books

People have been talking for years about gizmos that would take a book's contents as a cartridge or a computer file, and then display those contents on a screen good enough so that you could read the text as easily as you could read a book. Devices called the Rocketbook by NuvoMedia and the Softbook from Softbook entered the market, and then got gobbled up. Both Softbook and NuvoMedia were purchased by Gemstar (www.gemstar-ebook.com) in January 2000. There's virtually no remaining sign of the Softbook, and all that's left of the Rocketbook is the letter "R " in the book-reading devices Gemstar sells. They are the REB-1100 and REB-1200, $300 and $600 respectively. Presumbably REB stands for "Rocket Electronic Book." Franklin Electronic Publishers (www.franklin.com/ebookman/) sells similar—though less elaborate—devices called the Ebookman for about $140 to $200, depending on the model.

Many of these devices still have mutually incompatible operating systems and file formats. Until such time as the standards settle down regarding formats for such e-book readers, buyers run the risk of getting stuck with readers and text files in the equivalent of the eight-track or beta videotape format. The work-around to this seems to be that savvy e-book publishers are releasing their titles in multiple formats.

E-books are neat gadgets, but as of yet they can't quite compete with the convenience and reliability of "real" books. You can't easily read in the tub with such devices. Paperback books rarely have their batteries run out, or fall prey to computer viruses, or suffer crashes of their operating systems. You're allowed to read hardcover books during take-off and landing on most aircraft.

There are useful applications for e-book readers. Law students and medical students can store all their texts in one gadget, and have the text searchable. Travelers can carry a lot of reading material and still travel light. Readers with poor vision and partially sighted readers can increase font size for improved legibility. Husbands and wives can read in bed, in the dark, after their spouses have gone to sleep. But these are niche applications.

Not least of the problems is the expense. There are various

promotional come-ons, but at the end of the day, it will cost you over a hundred dollars, and maybe about six hundred, to buy an e-book reading device. Prices are going to have to fall a lot before these neat gadgets are a real threat to the paperback book.

The one real bright spot, so far as I am aware, is the discovery that personal digital assistants (PDAs) make almost ideal e-book reading devices. People are buying PDAs anyway, and then using them as e-book readers. For about the same money as a e-book reading device that can only do one thing, you can get a PDA that can do that job, and many other things besides. Much of the headway being made on e-books seems to be in that market, rather than in sales to purpose-built e-book readers, such as the Gemstar and Franklin readers.

There was a bit of a gold-rush mentality to e-books for a while there, but even the biggest cheerleaders have gone quiet. Stephen King's gimmicky publicity stunt back in March 2000 basically degenerated into otherwise more-or-less rational booksellers paying King for the right to give away his book.

Other e-book publicity stunts turned out to be just that—stunts. Promotion is all very well, but the industry being promoted doesn't really exist just yet. The actual installed base of e-book reading devices in use is still distinctly limited. While there are a fair number of early-adopters out there, there is as yet no real mass market demand for these devices or for the texts to go into them, and the early hype wasn't enough to sustain things. Even King's second attempt at e-book publication got nowhere. Since then, I haven't heard about any way for an author to make any *meaningful* amount of money from the sale of e-books.

However, if what you care about is wide distribution, and you aren't out to make a mint, and don't care that the book won't look exactly like a book, posting your book on the web for download could be the way to go.

The day may dawn when the e-book conquers publishing, but there are a few problems to be sorted out before that happens. So far, the results are not impressive.

Here's a quick list of additional websites regarding e-books.

This list is by no means exhaustive, but merely intended to give a quick snapshot of what's going on. As the subject is peripheral to the main subject of this book, I'm not to going to add all these companies to the *Names and Numbers* section. See the web sites listed below for more information. And just for the record, enough companies have bitten the dust, or been gobbled up, to produce close to a 50 percent turnover in this list since the last edition.

Ebook Websites	www.hiebook.com
www.overdrive.com	www.ebookmall.com
www.peanutpress.com	www.planetebook.com
www.rosettabooks.com	www.ebookstore.com
www.embiid.net	www.openebook.org
www.fictionwise.com	www.adobe.com/epaper/ebooks

12. Self-Published Book-On-Demand

That is to say, the subject of this book, spot-welded to self-publishing as described above. The creator of a self-published book-on-demand book writes, designs, prints out, binds, advertises, promotes, sells and ships the books in question. As it's what the rest of this book is about, we don't have to go into great detail here.

That just about wraps up our tour of publishing possibilities. What it boils down to is this: there are lots of jobs involved in being a publisher, but you can hire someone to do just about any part of the job for you. There is no single right way to do everything. Variables such as price, schedule, product quality, speed, money, storage space, work-time available, ego, demand for the book, and half a dozen others must be factored in to any calculation of the optimal way to get a given project done.

But we must also return to the question of whether the job should be done at all. *Opportunity cost* is the price you pay in lost opportunity whenever you choose to do anything. Spend ten bucks on lunch, and you have lost the opportunity to invest that ten dollars in the stock market. Spend a week, or a year, and fifty bucks, or five thousand dollars, on a book project, and whatever other opportunities that time and money represented are gone forever. So spend your time, talent, and money on things you really care about. If you've read this far, making books is likely one of them.

Making books yourself is fun and satisfying, and, if it is going well, selling them (or giving them away to appreciative recipients) can be as much fun as winning at Monopoly. Once you have the knack of it, book-on-demand could well open the doors to all sorts of pleasant, and maybe even profitable, endeavours.

So let's leave the business issues of publishing to one side for a while, and explore the mechanics of making your own books.

Chapter Three
Getting Started

1. The Basic Tasks

If you already have a decent computer and printer, you are within striking distance of being able to bind modest numbers of your own books. There is only one essential piece of equipment you lack: a heavy-duty paper-cutter. And, if you are willing to make some fairly heavy compromises insofar as how fast you work, and also about certain quality issues, you *might* be able to avoid the cutter. The minimum cost for a new cutter is about $700. Beyond that, you'll need perhaps a hundred or two hundred dollars worth of supplies and smaller bits of equipment. Certainly well under two thousand dollars would cover the initial investment and most or all of the initial operating expenses for even a fairly elaborate set-up. (See Appendix Two, *Sample Equipment Lists.*) For short runs, or one-off projects, there are unquestionably ways of avoiding most or all of that initial outlay, mostly by renting, buying, begging or borrowing the use of someone else's equipment.

Of course, you could easily spend just about any amount, and spending more could turn out to be worthwhile, if it improved your quality and speed of production.

Your material cost per book will be on the order of $1.00 to $6.00 per book, depending on the length of the text (and thus the number of pages), the cost of your covers, and so on. There is a very close relationship between the number of pages in of your book and how much it will cost you to produce it.

It will take you time to write or acquire the contents of the book, and time to lay the book out. You could easily spend hundreds of hours, and hundreds or thousands of dollars, just to get to the point where you're ready to *spend* that $6.00 per unit to make the actual books.

Nor are your first efforts likely to look precisely like "real" books. Your covers will be clunky, or the margins will be not quite right, or the spines will crackle alarmingly when opened. Pages might fall out. Three-quarters of the way through your print run you will discover a glaring typo. The odds are good that no self-respecting book shop would be willing to put your first attempts on their shelves.

Learning the skills of book-on-demand binding is a trial-and-error process, with early emphasis on "error." Don't plan on your first efforts being ready for the presentation case. But the basics are not all that tough, and the job of making books is, assuming you have reasonable mechanical ability, one you can do. Let's break that job down into its component parts. This list will have some overlap with the discussion of basic publishing tasks discussed in the last chapter, but is more detailed, and more focused on the mechanics of book-making.

Most of these steps we'll look at only in passing, as they are fairly straightforward, and because many of these tasks won't apply to all book-on-demand projects. But let's take a least a quick glance at all the basic tasks and consider them one by one.

- *Creating or obtaining the book's contents*. Obviously, before you can print the book, someone has to write the words and draw the pictures. Furthermore, you have to make sure you have the right to print and distribute the book. Even if you do have the legal right, you should make sure you aren't making trouble for yourself. I sometimes do book-on-demand versions of the books I write for commercial publication, in the form of Advance Reading Copies (ARCs) But before I do so, I make sure to get clearance from the contracted publisher. Whether or not a contract allows me to do ARCs can be a grey area. But if it's going to annoy the publisher, I lay off. Whatever good I could do the book will be more than offset by the damage I would do by angering my editor.

  If you're printing someone else's book, don't rely on a handshake deal. Write up a clear and specific contract that covers times and amounts of payment, confirms your right to publish and the author's right to the book's copyright, settles other rights and obligations on both sides, and that covers any likely eventualities. To emphasize, the contract should require a statement from the putative author that he or she is indeed the author and does indeed control the copyright. There are various sources for model contracts that will have standard boilerplate phrasing to cover most of this. Many writers' organizations publish such model contracts.

- *Interior design and layout.* The text and pictures need to be arranged on the book's pages. Design elements, such as running headers and footers, chapter titles, tables of contents, and so on, need to be set. Everything needs to be checked over and over again. The further along one is in a project, the harder it is to correct mistakes. For this reason, double-and triple-checking in the early stages is a tremendously important time saver.

- *Cover design and layout.* This is a different enough job from interior design that it needs to be considered separately. Spine lettering, cover stock, cover blurbs, ISBNs and barcodes (both close to essential if one wants to sell the book through a third party), artwork, and so on need to be considered. The cover is an advertisement for the book, and needs to be both effective and appropriate.

- *Prepublication marketing and promotion.* This consists of a whole series of issues that are largely outside the purview of this guide. If you want promotional work or material done for the book, it should be timed to help the book, and not happen too soon or too late. (That might sound blindingly obvious, but I've been at plenty of book events where everything but the book was there.) If you identify a particular market for the book, you should make sure those people know about the book in time, and you should do things that will make them want to buy the books: Get a cover blurb from someone they know, or have a flyer mailed to them explaining why this book is what they want. If there is an event that might be a good venue for selling the book, make sure to be there, and make sure the books will be ready in time. See Dan Poynter's *Self-Publishing Manual* for more information.

- *Prepublication legal and filing work.* If there is any quoted material in the book, copyright issues must be resolved, and permissions obtained. If you want any bookstore or mail order operation to sell your book, it must have an International Standard Book Number, or ISBN. In the United States, ISBNs are handled, not by the Library of Congress, as many think, but by the R.R. Bowker company. They put out the *Books In Print* books and CD-ROMs used by bookstores for ordering. The average bookstore can't order a book without an ISBN, any more than you could place a phone call to someone who didn't have a phone number. Without an ISBN, you're locked out of virtually every mainstream bookseller's shelves. As of this writing, a starter set of ten ISBNs costs $225, and that does not include barcoding. Contact R.R. Bowker early on, well before your publication date. Don't get behind on this, and wind up having your ISBN registered six months after the book comes out.

- *Printing of interior pages.* For the average home office, printing out one copy of an entire book-length manuscript can be a struggle, with paper jams, paper misfeeds, typos discovered only as the paper leaves the printer, and so on. Printing out a book double-sided, in a book-style layout, and printing out multiple copies of it, can easily make it more of a struggle. Good planning and good procedures can eliminate most (but not all) of these problems.

- *Printing the cover*, or having the printing done elsewhere. Printing the cover is the one job wherein it is most likely (but by no means certain) that you will want to have done by an outside print shop. However, color printers for the home office are getting good enough and cheap enough that it's getting much more practical to do it yourself. You will almost certainly want some color (by means of colored stock or color printing) on the cover, and will want it printed on somewhat oversized paper, and on a heavier cover-stock. A good cover might well cost as much to produce as the interior pages. But a cheesy-looking cover can make the book look bad, and actually be a *dis*incentive to picking up the book. No matter how cheap it is, a cover that drives readers away is no bargain. Don't cut corners.

- *Scoring* is the best way to insure a sharp square crease on a sheet of paper. Most hand-binding processes and some machine binding processes will require the cover to be scored in order to form the folds between the spine and back cover, and the spine and front cover.

- *First paper cut.* Often, but not always, needed. If you are producing a book with pages of 8½ by 5½ inches or smaller, it makes sense to print two pages side-by-side on letter-sized paper, and to then cut the sheet in half. This book is sized 8½ by 5½ inches, less trim (see below). In similar fashion, books up to 8½ by 7 inches can be produced two-up on legal-size paper. If you are getting your paper-cutting done outside your workshop, getting this cut could add an extra trip to the print shop.

- *Binding.* It is the problem of binding paperback books with which we will chiefly concern ourselves in this guide. The would-be basement book-on-demand publisher has almost certainly had some experience with nearly all the other design and manufacturing tasks in question, or with some job similar enough to signify. It's the actual sticking together of pages between two covers that people don't know about. For the purposes of this guide, we will mainly limit ourselves to adhesive-bound, or "perfect" bound books. The term "perfect" is the name given to the sort of binding done on virtually all paperbacks. In reality, it is by no means a perfect means of binding books, but if it's good enough for 99% of commercial paperbacks, it's good enough for our purposes.

- *Final Trim.* This is, in my opinion, the stage that makes a book look like a book. Before the final trim, the pages are not perfectly aligned together, and the cover is not square with the pages. But with three swift slices of the paper cutter, all of these imperfections are cut away, and suddenly there is a finished book.

- *Post-manufacture processing and marketing.* In short, another whole list of things that will not be discussed much in this guide, but that you, the would-be publisher, will have to worry about sooner or later. Packaging, storing, inventorying, selling and shipping your books are things you need to think about from square one.

  Various post-manufacturing issues might well affect a lot of your earlier decisions. For example: if you are selling a book mail-order to a technical audience, you can probably get away with a much simpler and cheaper (and perhaps less professional-looking) cover design. After all, they won't see the cover until they have already have purchased the book. But if you want to sell it in the local bookstore, or in some other retail environment, that book has to look every bit as sharp and professional as all the other books on the shelf.

As you can see, you need to make a lot of choices before you start. Will you need an ISBN? A barcode? What size book will best fit onto the store shelves? (A larger paperback might be out of luck on a shelf designed for mass-market paperbacks.) Will the books fit into the stock of shipping envelopes you just bought? Maybe by dropping the point-size on the type from 12 to 11 you can reduce the page count enough to make for cheaper printing or postage. (I made exactly that change to this guide for exactly that reason.)

There are lots of things to think about besides printing words on paper and gluing pages together. But none of the other issues matter unless or until you have printed those pages, and printed that cover, and glued them together, and trimmed the whole thing down into an honest-to-God book.

So let's consider the jobs of printing and binding, and the machines and gadgets that let you do them. In the pages that follow, we'll examine the machines, aside from binding equipment, you'll need to do the job. You'll get enough detail to get you working, although we're not going to be at the level of step-by-step instructions.

2. Equipment for Printing and Paper Handling

Small-scale book manufacture can be broken down into two basic categories: books bound completely by hand, and books bound using some sort of machinery more complex than clamps and paintbrushes. In later chapters, we'll discuss the various ways and means of binding, and machines for binding, in great detail.

In the remainder of this chapter we'll discuss all the other pieces of equipment, and some materials, needed for book-on-demand. Though the binding technique changes, depending on how big your operation is, everything else—how the books are laid out, printed, cut to size, and so on—stays more or less the same. No matter how you bind your books, you're going to print the pages on the same sort of printer. In this next section, we'll discuss the hardware you'll need no matter what sort of binding you do.

Monochrome (Black) Laser Printers

Anyone who want to do serious book-on-demand needs a big, fast, bruiser of a printer, and should have a duplexing printer. (A *duplex* or *duplexing* printer can print on both sides of a page in one pass through the printer.) Get one. At the very least, be sure that any new laser printer you buy has an available duplexing option you can add later. Even at very low production rates, a duplex printer makes sense.

Printer models and prices change fast. This year's top-rated new units are next year's bargain basement specials. There is no whatever I report about the latest printers would be dated in a matter of months, or maybe mere weeks. The key point is that printers are not just getting better, they're getting cheaper. Current models are often ten to twenty percent cheaper than the equivalent models of a year or two back, and they're faster and have more features.

Instead, I'll talk about what to look for in a printer intended for book-on-demand use. The key, as I have noted before, is to get a printer that can do duplexing. As we'll see in Chapter Four, it is possible to do book-on-demand with a printer that will only do single-sided printing, but you can bank on plenty of headaches and wasted time and effort.

Modern printers use something called "a page description language" that allows them to receive and interpret instructions on how to draw images on the page. Hewlett-Packard's page description language for its laser printers and modern ink-jets is "Page Control Language," or PCL. The latest version is PCL-6, though PCL-5 is far more commonly used as of this writing. Virtually all lasers use some version or emulation of PCL.

The other major page description language is PostScript. Most higher-end printers will have both PCL and PostScript. Some of the very fanciest might not have PCL, but will have PostScript. As we shall see, PostScript is a good thing to have.

The page description languages are backward compatible: A PostScript 3 printer can handle PostScript version 1 input, and a PCL-6 printer can deal with PCL-2. Get a printer that can handle both PCL-5 and PostScript Level 2 output. PCL-6 and PostScript 3 would be nice.

Many printers are designed to be upgradeable, so that, for example, you could plug in a PostScript 3 chip later on. Make sure your printer can be upgraded.

Get a printer that can do 600 dpi or better. (The abbreviation *dpi* is short for *dots per inch*, the basic measure of printer resolution or image sharpness). Older printers will only be capable of 300 dpi, which is acceptable for many purposes, but not really good enough for on-demand books. These days, it's getting hard to find laser printers that won't do 600 dpi, and 1200 is getting more common.

You want a printer that will stand up to relatively heavy use. Printers are often rated at pages-per-month, or some comparable measure. The average home laser printer might do 500 pages a month. You might run that many pages in a hour, and keep it up, hour after hour, day after day, if you're printing books. A light-duty printer intended for the average home user will soon put all four feet in the air and die under that sort of abuse.

Consider choosing last year's model of printer. It will likely be going cheap. However, accessories for older models of printer tend to vanish from the market before the printers themselves. For that reason, you'll probably want to buy whatever bells and whistles you want for the printer at the same time you get the printer itself. (The good news is that such accessories often surface on Ebay. If your printer needs a model 234 denobulator, the odds are that someone is trying to sell one on Ebay.)

All the laser printers on the market will run letter-size paper, and just about all will run legal size. Some printers, but not all, have adjustable paper trays that allow you to run odd-sized papers. You might well want to run custom-sized paper, perhaps 6 inch by 9 inch, or 8½ by 5½ inch. Go with a printer that lets you do this. It could come in very handy. Read Chapter Four, *Page Design and Printing*, to find out why. There are some printers that will allow you to run 11 by 17 (tabloid) paper, or even larger, and still do duplex printing. These machines run into money, as you might imagine. Most book-on-demand printers don't *need* to run pages that big, but the capacity could turn out to be useful. There are occasions where you might want to want to run letter-sized paper sideways, with the wide side of the paper facing the print path. A printer that can handle 11 by 17 can do this—and can also print out flyers, posters, fold-down mailers, etc.

When you're shopping for a printer, take a good hard look at toner-cost per page: the standard measure is 5% coverage, which means, more or less, a double-spaced typed page. More than likely, your book pages will have more than 5% coverage. My rough estimate is that the print on this book takes up 8% of the page, though that might be low.

There's a lot of blue smoke and mirrors in the claims regarding toner-cost-per-page made by the printer manufacturers. The manufacturers base their estimates on a "standard" page without much text on it, an assumption that drives the price per page down. However, they also assume that you'll pay full retail for their cartridges, that you won't buy cheaper third-party cartridges, and that you won't reload your cartridges, all of which drives the price per page back up. They then divide one unrealistic number into the other in order to get their estimate of your cost. However, nobody pays retail, and your pages won't match their carefully designed test pages anyway. In other words, you're on your own working this one out.

Spend some time on the question. Cost of toner cartridges is a significant issue: buy five or ten cartridges for certain machines, and you've pretty much spent what the printer itself cost.

There are lots of companies that sell off-brand or reloaded or remanufactured cartridges for about half or two-thirds the cost of new, name-brand cartridges. See what they're offering for the printer you have in mind. Using reloaded cartridges might void your printer's warranty, but the reloads are a lot cheaper. Buy your reload from a reputable firm and they shouldn't do your printer any actual harm. Image quality *might* be as good as from a name brand cartridge, or it might be a little off. You pays your money and you takes your choice.

One other point currently in issue regarding toner cartridges: some manufacturers (Lexmark, I regret to say, is leading the way on this one) are starting to put special electronic chips on their toner cartridges. These chips are hooked into the toner-level sensor. This is the gizmo that tells the printer how much toner it has left. Once it gets to the toner-out position, the special chip essentially locks the sensor down into the toner-out position and leaves it there permanently. This means that you could refill the cartridge until you were blue in the face, but it would still show as empty, and the printer would refuse to run. This is the point where the fun really begins. The cartridge reloaders then came up with their own chips that behaved like the lock-down chips, *but* allowed reloading. Then Lexmark sued. Their theory, if I understand it, was that when someone bought a cartridge with such a chip, they weren't buying the right to fiddle with it or refill it. All they were allowed to do was give it back to Lexmark. Therefore, if a cartridge reloader took such an empty and reloaded it, that was naughty and the judge should make them stop. Or something. The legal arguments seemed a little lean to me, but the case has not yet been resolved as of this writing. *Recharger* Magazine has a series of articles concerning this case on their website.

I should note that while Lexmark seems to be taking the lead on this one, you can bet the other printer makers are watching carefully. After all, they all make a *lot* of their money selling toner cartridges. For a while there, it seemed as if the reloaders and remanufactures were staying well away from Lexmark cartridges for fear of being sued. As things currently stand, a few rulings have gone against Lexmark. The reloaders and remanufacturers seem to have their confidence back and are merrily slapping their anti-lockdown chips on cartridges and selling them. But of course, that could change.

The moral of the story is that you should make sure any printer you buy doesn't require cartridges with lock-down (aka "killer") chips—or else make sure that someone else has figured out a way to defeat the killer chip.

I currently use the Lexmark Optra T614nl with duplexer (and remanufactured cartridge complete with anti-killer chip) for my page printing, and am quite pleased with it. My previous printer, a Lexmark 4039Plus, still runs just fine, but it didn't have the speed or paper handling I needed plus it wouldn't duplex odd-sized paper. It also didn't run short-grain paper well. The T614 does all those things. As we'll see in the section on paper grain, this can be important.

Color Laser Printers

Book covers should have color. A cover printed in black on white paper just doesn't look professional. You can get color by using colored stock (which rarely looks professional) or by using either a color laser or color ink-jet printer.

If you want to do wrap-around covers (as on a paperback book) you will need a printer that can run paper as wide as your book page is tall, and as long as twice the width of your book page plus the width of the spine. For an 8½ inch by 5½ inch book that's an inch thick, that comes to a cover 8½ by 12 inches. Probably you'd just print it on legal-size paper (8½ by 14) and trim the excess. For a 6 by 9 inch book a half inch thick, you'd need a cover 9 inches by 12½, which is a little on the oddball side. You'd have to cut down something from larger stock. Usually it's easier to print the cove first, and then cut the paper down to size for binding. This also lets you print something on the part of the stock that won't be the cover—a postcard, a bookmark, or some other promotional piece, for example.

Color laser printers are getting massively cheaper, but prices still start at about $1,000. Cost per full-color, 100% cover page is about 33¢ to 53¢, which is comparable to the cost of doing the job with an ink-jet. I have seen new color printers that can do 11 x 17 color offered for sale at $2000.

I should mention the Tektronic "solid ink" printers sold by Xerox. These are not laser printers. They use, not toner, but blocks of colored resin, to put an image on paper. They are by all accounts high-quality printers, and sell at attractive prices. However, I have no information concerning the durability of the images they produce, nor do I know how well they laminate, or how well these printers would do at printing covers. I'll try and investigate this point for a later edition of the book. For the time being, you're on your own.

Note, however, that an image printed with color toner, like one printed with black toner, is, generally speaking, too fragile to stand up to the rough-and-tumble treatment a paperback book cover endures. You will need to use special cover stock or else will need to protect the cover with some sort of laminate or overcoat. These issues are discussed in detail later in this chapter.

Toner images do have the advantage of not being water soluble, the way most ink-jet images are. A cover printed on a color laser won't dissolve if held by someone with sweaty hands.

Ink-Jet Printers

Ink-jet printers are evolving even faster than laser printers, so there's even less point in discussing specific models. However, there are features to put on your shopping list. Make sure that your printer is a four-color printer that uses true black ink for black on the page. Three-color printers mix the three colored inks to produce a muddy grey for black. (Three-color printers are becoming rarer and rarer, fortunately.) Make sure that, at the very least, the black ink is in a separate cartridge, or better still, that all the ink colors are in separate cartridges. If all the inks are in one cartridge, and one color runs out before the others, then you have to throw out a lot of perfectly good ink in order to swap cartridges. Be sure to consider carefully how much the cartridges hold, and how much they cost, and how easy they are to find. Also look into what's involved in reloading the cartridges for the printer you are considered. There are kits to refill most of the ink cartridges out there, but be sure to wear old clothes and rubber gloves when you do the refilling.

For some reason, (probably the reason involves lawyers) it seems as if there are more refill options and third-party cartridges available for Canon and Epson printers than there for HP printers, and the HP refilling systems are often more ungainly, requiring you to refill the ink by hand.

Make sure the printer can handle the paper size you need. For the covers on 8½ by 5½ inch (pre-trim size) books, you'll want to be able to handle at least legal size, 8½ by 14 inch paper. Nearly all the printers can deal with that. For larger books, you'll need a printer than can deal with 11 by 17 inch paper. Printers that can handle 11 by 17 ink-jets are not all that common, but they are available, and reasonably priced.

Also make sure your printer can handle heavy cover stock smoothly and accurately. It's not enough for it to handle cover stock fairly well: you need consistency and reliability.

All other things being equal, the straighter the paper path, the easier it is for the paper to feed thick stock. Many Hewlett-Packard ink jet printers use a U-shaped paper path. HP makes excellent printers, but test the paper feed for cover stock carefully before buying one for cover printing.

Avoid the low-cost "Windows Only" printers. These machines don't have any real on-board printing logic, and instead rely completely on Windows for printer control. Because they are relying on the computer to do their thinking, this trick can make the printer pig-slow, and also slow up your computer during print operations. (And if you're not running Windows, you're out of luck. The printer might not work at all with any other operating system.)

Most current color ink-jet and laser printers, even the ones that aren't "Windows Only," do rely on the computer to do a fair amount of their image processing and spooling. What this means is that the printer's speed is highly dependent on the computer's speed, memory size, and hard drive size. So if you want your ink-jet printer to go faster, beef up your computer.

As with laser printers, ink-jet printers come and go fast. If an older model is on sale cheap, and it does what you want, that might be a good way to save money—but make sure the cartridges for that model are still available before you buy.

The current color printers, laser and ink-jet, do a good job of printing color accurately, but they aren't going to be utterly precise. Getting *perfect* color is going to be a lot harder than getting *good* color. I will freely admit that I am a babe in the woods when it comes to CMS, PMS, Pantone, RBG, CMYK, and all the other alphabet soup of color printing. What I have learned is that what comes off a color printer at home is not likely to have anything more than a family resemblance to what the same input file would produce at a professional print shop. High quality color printing is, by all accounts, a bear.

For my printed-at-home covers, I use the Hewlett-Packard 1120c, which claims 600 dpi-quality output, does nice sharp color, and handles 11 by 17 (in fact, 13 by 19) paper. It has a straight-through manual-feed slot on the back which handles heavy stock, as the U-shaped paper path is a nuisance at times with certain paper stocks. It cost me $500. The price dropped to $400, and then the 1120c was discontinued in favor of this year's model. I have had only fair luck with refilled cartridges, and general stick with new HP cartridges. The 1120c is a bit slow, but it prints out one cover faster than my laser prints out one book, which I would say is fast enough. There might be faster, higher-resolution, printers out there, and I wish it had separate ink cartridges for the colors, but still it does the job.

Paper Cutters (1) For The Initial Cut

If you're going to make paperback books, you're going to have to cut paper. You'll need to cut through a height of paper as thick as the whole book, and do it three or four times for each book you make—and you'll need to do it with a high degree of accuracy. You will also likely need to cut down oversize stock fairly often, and do it with great precision.

Many book-on-demand printers will also need to cut the pages apart before they are bound. This cut doesn't *need* to be done on a heavy-duty cutter. You *could* cut ten or twenty sheets at a time of your 330-page book on an office paper trimmer. But if you were doing more than two or three copies of such a book, it would become vastly inefficient. For anything besides the smallest possible production rate, you need something that can cut through more than ten pages at a time.

Cutting paper is easy. But cutting an inch-thick stack of paper with sufficient accuracy, smoothness, repeatability and reliability takes a bit more doing. There is really only one good way to do it: with a guillotine paper cutter, a tool expressly designed for the job.

A guillotine cutter uses a steel blade anywhere from twelve to thirty inches wide. You position the adjustable backstop to whatever cutting length you want, square up the paper against the backstop, set a clamp to hold the paper in place, and then bring that big steel blade down through the paper.

Most guillotine cutters are electric or even hydraulic. Push a button and the blade goes down. Some use a manually operated lever, so you're using muscle power to do the cutting. Various safety features make sure you don't have any fingers in the way of the blade. My cutter, for example, won't cut unless two buttons, more then two feet from each other and well away from the blade, are pressed at the same time. It's impossible for your fingers to be under the blade while you're pushing the buttons.

The fancier models have all sorts of special features like memory presets and programmable cutting patterns, but such machines are way out of any home operation's price range.

The cheapest, most basic hand-operated guillotine cutter I have seen (a manual unit made by Martin-Yale) goes for somewhere in the neighborhood of $500-$900 when purchased new. It is sometimes available via special order at office supply outlets, as well as through *Quill Office Supplies* and the *Printers' Shopper* catalog, *MachineRunner.com*, *Factory-Express.com*, etc. It is possible to spend well over two or three thousand dollars on a modestly equipped electric cutter, and several thousand on one with all the options.

The *good* news is that, aside from sharpening or replacing the blade, cutters capable of dealing with an inch-high stack of paper last more or less forever. I bought a twenty-year-old electrically operated guillotine for $450 about ten years ago, and it has served me well with just about zero maintenance. The heaviest duty use I will ever put it to will never come close to what it was designed to handle.

Do you really need such a cutter? *Maybe* not, but I'd say yes. Rupert Evans, author of the extremely useful book *Book-On-Demand Publishing*, wrote an article in *Flash* magazine on ways to get cutting for book-on-demand done without a guillotine. He discussed techniques ranging from slicing a few pages at a time with an X-acto knife to running the book pages through a table saw. The photos that he used to illustrate his text, presumably intended to illustrate the best possible results, looked pretty bad to me. Even Evans didn't appear too convinced by his results.

However Dr. Evans does report one bookmaker who cuts his books "using a plywood blade on a cutoff saw. He uses it both for cutting sheets in half, and for trimming bound books. The cuts have a faint, very attractive pattern... caused by some sort of vibration. One would never guess that such a cut was made by a rotating blade." However, unless you have a cut-off saw already, that information isn't likely to be much use—and you might not find that patterning attractive on *your* book pages.

In short, unless you're willing to put up with some extremely severe restrictions on your book-making career, get a guillotine or get access to one.

There is one other alternative to a guillotine, a *lying press and plough*, but this system is so time consuming—and antiquated—that it's probably not a useful option. In most cases, the press and plough will likely be nearly as expensive as a cheaper guillotine. Any book on doing hand-bound hardcovers will discuss the lying press and plough in detail, but I'll touch on it briefly here for the sake of completeness.

The lying press is used for lots of processes in making hand-bound hardcover books. It is essentially a large vertical clamp, usually of wood. For cutting, the pages of a book are placed in it spine-side down.

The plough is a knife blade set in a carrier that slides up and down along the side of the press. The blade is moved down the length of the book, slicing one or two sheets off. The blade is then moved in very slightly toward the book by means of a screw mechanism, and the blade carrier is slid back to cut through the next page. The blade is then advanced in again, and the carrier slid forward to cut the next page. Repeat *ad infinitum*.

Besides being a very slow way to work, the plough is suited only to trimming away rough edges. It wouldn't do very well cutting a stack of paper in half.

There is one experiment in cutting thick stacks of paper that

I would like to make, but have not tried yet. It seems to me you could cut a book with a heavy-duty matte cutter, designed to slice through ¼-inch matte board. If you had a good steady cutting guide, and your paper was well clamped down, and you took enough care, it seems to me a matte cutter might do the trick in two or three passes, more or less—if the book weren't too thick.

There is one other possible, partial solution to the problem. Most book-on-demand books will be half-letter size, or 8½ by 5½ inches. *Imation* makes a interesting line of laser-perfed paper stock, including a "booklet bond" that is letter-sized paper perforated to split into half-letter sized sheets. Last time I checked, this paper was pricey, and only available in larger quantities. However, Quill sells a similar paper in smaller amounts for a reasonable price. But even with such paper stock, you'd either have to settle for books that weren't trimmed square after binding, or else you would have to come up with some way to do the trimming by hand. (See the discussion of Chet Novicki's "Gigabooks" technique at the end of Chapter Six for other ways to avoid paper-cutting.)

To sum up, for any sort of real production work, there's no getting around the need for a proper guillotine cutter. You need something that can clamp down a whole stack of paper, holding it tightly in position, while a big steel blade slices straight through. That's what a guillotine was designed to do. An office paper trimmer can cut loose paper piecemeal, ten sheets at a time. However, a trimmer won't be able to cut through the whole bound book as is required for the final trim, as we'll see in the next section.

If you are only doing a one-off project, or only doing occasional work, or just doing a test run or two before deciding how heavily you want to invest in book-on-demand, there should be plenty of ways to buy or filch paper-cutting services. If you want to pay retail, print shops generally charge between 50¢ and $2.00 per cut. That can add up fast, and there is the nuisance of carrying your work back and forth to where the cutter is.

There are other ways. Poke your nose into the art department at your local community college and see if there's a cutter they'll let you use. Maybe you work days at an office big enough to have its own print shop. Maybe you have a friend who works at such a place. There are plenty of cutters out there. It's a question of getting at them without spending too much, and without causing pointless trouble.

Needless to say, such games will get very old very fast if you decide to go into book-on-demand in any serious way. The best solution is to buy yourself a new or used cutter. There are lots of places to look. I found mine in the classified ads of the local paper. *North East Printing Machinery Inc.*, *The Print Shopper*, and *TPX Online* are all out there on the World Wide Web, each offering some form of classified advertising for used or discount printing equipment. There all lots of other sites that come and go, offering similar services. And, of course, there's Ebay. See the *Names and Numbers* list at the back of this book for particulars.

Paper Cutters (2) For The Final Trim

The guillotine cutter comes into use again after the book is bound into its cover. It's used to trim away any excess paper, and to tidy things up. A paperback book is a rather scruffy-looking item until it is properly trimmed. The idea of the trim is not to remove a lot of paper, but to even things up and make them look sharp and professional. To me, it's this final trim that makes a book look like a book, rather than like a stack of paper that has been glued together.

Obviously, you need to plan for this final trim starting with your initial layout, by allowing extra-wide margins at the top, bottom, and outer edge of the book pages. Bear in mind that even the best guillotine (and it's doubtful a basement book-on-demand operation has the best) is not always one hundred percent accurate. The Martin-Yale manual guillotine mentioned above doesn't even claim an accuracy better than 1/32 of a inch, and that's assuming that the operator lines up every cut perfectly every time. You have to assume that you're going to be off by a sixteenth of an inch now and again. If you've only allowed yourself a quarter-inch margin, you are quite literally going to be cutting it close.

Herewith, a few obvious and not-so obvious tips that will make your life as a paper-cutter that much easier.

If you are printing ten copies of a book, don't do all three cuts on each book at once. Instead, for example, do the tops of the ten books one after the other, then the bottoms, then the outer edges. This is not only more efficient, it will make your books more consistent. You might find a different order more efficient, but find whatever works for you and then always do the three cuts in the same order, likewise to encourage consistency in the final appearance of the book.

Follow consistent procedure throughout, but before you set those procedures, experiment and find the rules that work for you. Take note of how your books behave in the cutter, and adapt your procedures accordingly. You might find that your cutter does better if you make all cuts with the book face up, or all cuts with the book face down. Or perhaps you'll find your cutter works best if you make all the side cuts with the bind against the paper guide, or opposite the guide. Probably you will discover some small advantage to one particular choice in such matters, depending on your equipment. My books tend to curl toward the front of the book a bit after binding (though they usually flatten out after a while). I cut the books face-down as much as possible, so that gravity and the pressure of the cutter clamp are working together to flatten the book during the cut.

If you always do everything one way, your books stand a better chance of having a matched appearance. You don't want ten copies of your book to be ten different sizes when you put them in a stack. (This is yet another of the items that go in the categories of things I learned from my mistakes.)

Finally, here is possibly the biggest headache-eliminator I derived from Rupert Evans' book. Thanks to the adhesive, the spine edge of a glued book is thicker than the rest of the book.

This matters because guillotine cutters use a clamp to hold the paper in position while cutting.

The clamp can come down harder on the spine than the rest of the book, crushing it and making the book unsalable. Put a spacer between the book and the clamp, making sure to keep the spacer well away from the spine. This will save many a book from being ruined at the last stage of manufacture. Evans suggests using chip board or the binding board used for hardcover books for the spacer, and gluing the spacer to flexible magnetic material. You can slap such a magnetic spacer into position and it will stay attached to the cutter clamp (assuming your clamp is steel or iron), and yet out of the way.

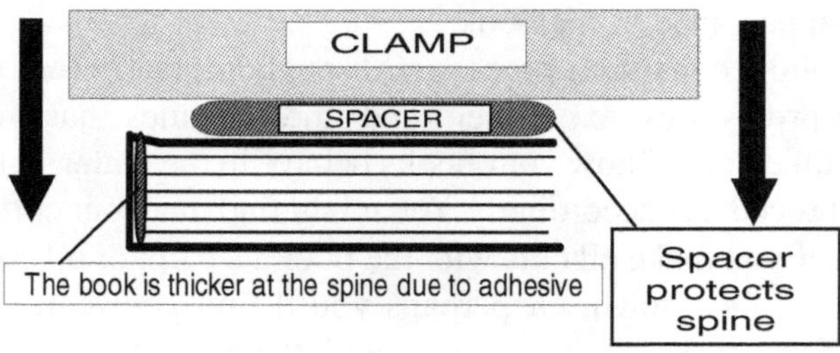
CLAMP
SPACER
The book is thicker at the spine due to adhesive
Spacer protects spine

3M makes a purpose-made magnetic clamp pad, and using that pad very definitely improved the precision and squareness of the final trim on my books. *Printers Shopper* sells these pads. Or else, as Dr. Evans suggests, you could save yourself $25 or so by taking a bunch of the freebie magnets that end up on your refrigerator, and gluing them to chip board with contact cement.

Scoring Tools & Techniques

Scoring is based on a pretty simple concept: paper folds better when it has a nice sharp crease. You want your paperback books to have that nice sharp crease on either side of the spine.

Some binding processes—mostly the ones involving expensive machinery—don't require scoring. But if you're doing handbinding, and you want paperback books with wrap-around covers, you're going to have to score your covers.

Scoring ought to be a no-brainer, but it is quite possible—even easy—to get it wrong. Scoring needs to be done with a fair degree of precision.

There are ways to get fancy with it, but the basic idea is quite simple: Measure the precise width of the stack of pages that will make up your book's interior pages. (Get a caliper with a nice, easier-to-read display for this. Don't hold a ruler up against the paper stack and squint.) Working on the inside of your printed cover, mark on either side of the center line, exactly half the width of your pages on either side. In other words, if you have an inch-thick book, mark the spine half an inch either side of the spine center. You will probably need to allow for a little extra width, to allow room for the binding adhesive to slosh out a bit from the spine. How much, and whether, you will need to do this, will of course depend on your binding procedure. It will require a bit of experimentation to get the fit right.

Use a T-square or some other sort of straight edge to guide you, and score the paper. Rupert Evans uses a ballpoint pen drawn with steady pressure, using a ruler to guide it. (The final bind will hide the ink.) That all that's needed to put a sufficient bend or crease into the paper. It works. I have also done well using a *very* dull old X-Acto blade. Whatever implement you use, don't press too hard. Remember you're not trying to cut through the cover. You're just encouraging it to fold.

I've read here and there that it's considered better to score paper on the outside of the cover. However, that's very tough to do with the above forms of manual scoring without damaging the printing on the cover. Scoring the interior of the pages has always worked for me, and I feel gives a better appearance to the final book.

The above techniques will do for very short print runs. But if you're doing any significant amount of handwork, a scoring machine as discussed below will help insure consistent, repeatable results, and cut down on the amount of time and effort needed to align and position the cover stock for each score. It will also cut down on the number of covers ruined by bad scoring.

The first class of scoring machines are manually-operated. *Printer's Shopper* offers a few, ranging from about $75 to $350 dollars. These machines are basically rotary paper trimmers with special blades, and a few bells and whistles. In fact, some rotary trimmers can take scoring blades, and perforating blades as well.

There are two problem with these machines. First, they can only score one line at a time. Most binding techniques that require scoring want two scores per cover—and some want four. The second is the amount of fiddly measurement required to set each scoring position. There are, of course, ways to deal with these problems—just apply money.

The above methods, sliding a blade along a straight edge, or sliding a round blade held by a guide arm, rely on a small object dragging down the whole width of the paper. The other way to go is with what you might call a bar scorer. On this type of machine, a thin blade, as long or longer than the paper is wide, is pressed down across the whole width of the paper in one go. The bar is pressed down by pulling on a manually operated lever, or by pushing an electric switch. Most of the bar scoring machines have two paper stops. Once you have your measurement set, you can move from the first scoring position to the second without resetting the measurements.

A number of companies, including *Standard*, *Maping*, and *Powis Parker* market these machines. They range in price from about $400 to $1,700(!) and probably on up from there.

The Powis Parker unit, simply called the Scoring Machine, goes for $875. That's a pretty stiff piece of change, but it does have some nifty features. First off, the alignment bar is transparent, so you see the spine of the book as you're aligning things. This allows you to do a good eyeball measurement and centering of the spine lettering. Even better, it has a clever mechanism that reads the width of the book directly. Place the book pages in a measuring gauge, slide the clamp tight against it, and the Scoring Machine "knows" exactly how wide your book is.

Sloppy scoring makes for books with sloppy spines. A book

with a sloppy, un-square, or crooked spine looks awful. Do everything you can to get the scoring right. Work out a nice repeatable procedure. If practical, consider printing scoring guides on the inside of the cover, in exactly the right position for your scoring. Obviously, it is crucial that these guide lines be well-aligned with the printing on the *front* of your cover.

Bear in mind that paper has a grain, just like wood. As with wood, cutting parallel to the grain will result in a cleaner cut than going against the grain. So too with scoring. Try and arrange matters so that your book's spine runs parallel to the paper's grain. The packaging will usually indicate whether the paper is *long* or *short* grain. (Long grain has the grain parallel to the long side of the sheet. See the discussion of paper grain in Chapter Four.) However, it can be tough to find short-grain paper stock in the proper sizes for a book cover. At present, I buy 22 by 17 inch long-grain stock, and cut it down to 17 by 11 short-grain stock, on which I print my covers.

Throughout the binding process, work to make sure the folded cover and the assembled book pages, and everything else, are as squared-up as possible. Uneven stack of pages, or crooked spines, don't just look bad. They can cause the pages and covers to skew about during the final trim, resulting in cockeyed pages and covers and weak binds.

Likewise, do everything you can to make sure the pages fit snugly and precisely into the scored cover. If the scores are too far apart, making the spine too thick, the pages will not support the spine, and the spine will get crushed. If the scores are too close, and the spine too thin, more than likely the front and back pages will tend to drift up away from the spine during binding. This produces a weak bind that can cause pages to fall out.

Don't try to make a incorrectly scored and folded cover fit. Efforts to repair it will almost certainly do little more than make things worse. An unsquare, poorly scored, or inaccurately scored cover draws attention to itself, and makes the rest of the book look bad and lowers the quality of the bind. Discard any misscored cover and try again with a fresh copy.

Varnishes and Laminators

While toner or ink-jet output on normal paper is sturdy enough for interior pages, it is not durable enough for use on a book cover. The cover needs some sort of protection, either a varnish or lacquer of some sort, or else it needs to be laminated. Most commercial paperback book covers are varnished. Some commercial books have laminated covers.

Unprotected ink-jet ink can smear easily when exposed to the slightest bit of moisture. However, a light coat of Krylon UV resistant clear spray varnish does a nice job of protection against the peril of sweaty palms. There are lots of other similar spray-fixative products on the market that ought to do the job, though I wouldn't try soaking ink-jet printing in the tub, no matter what was sprayed on over the ink.

There are drawbacks to using the sprays. Spraying all your covers could get to be a cumbersome and fume-ridden job, and you'd go through a lot of spray cans. It would probably be a bit tricky laying down a consistent layer of spray-on varnish as well. There's also the question of tactile sensation. My Krylon-sprayed covers just felt a little odd to the touch. You don't want to end up with a book cover no one wants to hold.

You could probably rig up a pretty straightforward varnishing line with a paint roller or airbrush, or maybe by applying the varnish using a silk-screening frame. Probably any of these would lay down a thicker, stronger, and smoother layer of varnish than I got with the spray can. You'd have to rig up some sort of drying rack, and make sure you have a dust-free area and good ventilation. I have made only very limited experiments in this area, mainly because I wasn't crazy about the fumes. However, there's no reason varnished covers shouldn't work. Obviously, you'd have to test various products to come up with one that worked for you.

Let's turn to the other alternative, lamination. There are lots of laminating machines out there, but most are too expensive, too slow, or set down a laminate that is far too thick for use in a book cover.

Most laminators use heat to melt a sheet of clear plastic to a sheet of material, usually but not always paper. In double-sided thermal lamination, the material to be laminated is sandwiched between the layers of laminate and then run through a pair of heated rollers which use heat and pressure to bond everything together.

However, single-side lamination is to be preferred for book covers. "Real" paperback books from commercial publishers do not have varnish or laminate on the inside of their covers. Using a double-sided laminate will therefore make your book seem that much less like a conventional book. Laminating both sides of the cover can also make the inside cover so slick and slippery that the interior pages will simply slip out from between your fingers and the cover as you hold the book. A double-sided laminate can also wind up making the cover too thick and stiff, making the book more difficult to handle. However, these are relatively minor objections, and basically come down to personal taste.

The one serious mechanical problem is that many adhesives won't work well with a laminated cover interior. Some thermal glues might be hot enough to melt the laminate and cause a mess. And many glues simply won't hold well enough. The pages might stick to the cover at first, but as the cover is flexed by being opened and shut, the cover could work its way loose, making the adhesive crack, weakening the bind until the cover simply falls off. Some laminates will work with some glues. Others won't. The only way to know about any given combination is to experiment.

But if your adhesive and laminate can get along, and if you like the way double-side laminated covers look, and if you can get double-sided lamination done cheaply, then go for it.

The trouble with most lamination systems is that either the laminating machine is reasonably cheap, but the laminating material is too pricey, or else the material is inexpensive, but is applied by a very expensive machine.

The smaller machines, available at office supply stores, use lamination pouches, which are basically flat clear plastic sleeves. These come in various sizes, with the letter and legal size pouches going for fifty cents or a dollar each. That rapidly becomes awfully expensive. They also produce a laminate that is far too thick for use as a book cover.

The bigger machines use long rolls of laminate, which come in much cheaper on a per-unit basis than the pouches, and lay down a thinner laminate, but prices for these machines start at about a thousand dollars and head straight up.

Single-sided lamination has one potential drawback. Because one side of the paper is sealed against moisture, but the other isn't, the covers can tend to curl.

The best sort of lamination for book covers is single-side nylon laminate. Single-side lamination leaves the interior of the book cover un-laminated. Because nylon does not (or at least isn't supposed to) curl in humid weather, covers with single-side laminated nylon covers shouldn't curl up in dank conditions.

GBC (the *General Binding Corporation*) sell a broad range of laminators—but the cheapest that does single-sided nylon lamination has a retail prince of $7,000. The unit is called the GBC Eagle OS 35. (The *OS* stands for "one-sided.") This unit is designed to cope with the silicon fuser oil that winds up on the surface of most things printed by superfast color laser printers, such as the DocuColor 70 or DocuColor 40. Most laminators don't bond properly to silicon oil, so laminations of material coming out of such printers tends to fall apart after a while. (As of this writing I am told that a smaller and cheaper one-sided laminator might be on the way from GBC.)

Banner American distributes the *Foliant* line of laminators, designed specifically to do single-sided lamination. The cheapest model, the 370, sells for about $5,000. The Foliant line is likewise designed to deal with fuser oils—provided you use the proper laminate.

An alternate would to be to use a double-sided laminator such as the plain old GBC model 35, without the "OS" features, and do double-sided lamination. The model 35 sells for about

$1,500. GBC supposedly makes a special "stampable, glueable" laminate for the 35 that is supposed to bond to thermal adhesive, but it's hard to track down information about the material.

Even using pricey special purpose laminate, if you use one of these big roll laminators, your material cost for laminate per cover shouldn't come to much more than about 5¢.

There is another option, one that seems to strike a happy medium on price and practicality. *Xyron* makes a cold-lamination machine called the Xyron 850. (Note: as of this writing, the model 900 seems to be replacing the 850, which is apparently being discontinued. However, 850s are generally available for the moment, and the supplies for it should stay around.) and larger machines called the Xyron 1200 and 2500. The 850 and 900 handle material up to 8½ inches wide, while the 1200 handles foot-wide material, and the 2500 handles material up to 25 inches wide. The operation of the machines is identical, aside from the width of material that can be laminated. All use rolls of laminate available in various lengths, ranging from 20 feet to 200 or more. The laminate is self-adhesive, and does not require heat or electricity to be applied. The material to be laminated is fed in one side, and the operator turns a manual crank which feeds the material through and lays down the laminate. The operator then uses a built-in cutter to cut the laminated material away.

Various forms of laminate are available, including double-sided and single-sided (of most interest to book-on-demand printing) and also a laminate on one side, with a peel-and-stick adhesive on the other side.

Pricing seems reasonable, though the range of prices quoted by various dealers I found by searching on "Xyron" on the Web was quite broad. The 850 and 900 go for somewhere around $100, depending on discount and what "starter set" of laminate is bundled in. The 1200 goes for about $150.00. The 2500 costs over $1,000. A hundred-foot roll of the single-sided laminate for the 850 goes for between $24-$30. A hundred feet would bind 85 legal-size (8½ by 14) covers. That comes to about 30¢ to 35¢ a cover. (Important note: as of this writing, it seems that single-sided laminate is available for the 850 and the 1200, but *not* the 900. Xyron has discontinued several models, and added new ones. Check the Xyron website for current information.)

I used the 850 for quite a while, and got acceptable to good results. However, I finally concluded that the single-sided laminate put too much persistant curl into my covers. The double-sided laminate did lay flat, but as we will discuss next, binding a book with the inside of the cover laminated can get tricky.

Because many book-on-demand binding procedures involve heating the cover, and because the covers would have to be laminated before binding, many binding techniques that apply heat through the spine of the cover might not work too well with this cold-lamination system. The heat could damage the laminate—or the heater. A gentleman at Xyron said he thought the material could resist heat to 270 degrees, and probably rather higher if the heat exposure was brief and gentle.

I am not sure how well the Xyron material would do on color-laser printed material coming off a printer that used silicon oil, as discussed above.

3M slaps their own label on the Xyron machines. The laminating materials for the 3M and Xyron systems are interchangeable. Pricing for the 3M version seems comparable. *Brother International* sells the *Cool Laminator* for about $200. It is similar to the Xyron, except it uses an electric drive instead of a hand-crank, and does not seem to offer a single-side laminate.

Regardless of the type of lamination you use, there is a way to get single-sided lamination out of a double-sided laminator. Put two covers through the machine, both covers face out. Trim away the laminate from around the edges, and you should have two single-sided laminated covers. This could save you money as well. For example, the double-sided Xyron 850 cartridges go for $40, while the single-sided cartridges go for $30. Do the math. See if the trick works for you.

Finally, there are manually-applied laminates. These are peel-and-stick material. Pull off the backing and apply them to whatever you want to cover. The great advantage of these materials is that they can be applied after the book is bound. This could come in handy if you bind your books via a process that could damage covers that were already laminated. Many types of such manual laminates, along with glues for binding, and many other products of interest, are available through the *Brodart* library supplies catalog. I have gotten my best laminating results with hand-applied cold laminate. However, it requires more hand-work to apply. At present, I use a 4-mil laminate out of the Brodart catalog. The type I buy comes with a peel-away backing that's split down the middle, so I can peel back one side at a time.

I laminate before binding, simply because it's easier to work with a flat piece of paper than a finished book. I cut the laminate to the pieces the height of the books I am making (usually 8½ or 9 inches, plus a little extra for trim). I use a shallow box-lid as an alignment guide. I put my cover, face up, into the box lid, and slide it up into a corner of the lid. I set the laminate down on top of it, adhesive side down, and use the corner of the lid to line it up with the cover. I hold down one side of the laminate with one hand, and use the other to peel the backing away, moving carefully from the center toward the edge, smoothing the laminate as I go. With one side stuck down, I no longer have to worry about alignment on the other side. I peel and pull the laminate away in the same manner, and then use a soft burnishing tool or a rubber roller to flatten down the laminate completely. It take a minute or two per cover, and costs about 5¢ a cover.

There's one other factor to bear in mind concerning the cost of lamination. If you laminate your covers, then the paper stock for the covers can be a lot lighter and less durable—and therefore cheaper—than it would have to be otherwise. My lamination pays for itself by allowing me to use lighter, cheaper paper stock.

Now that we have a basic understanding of the tools we need to make books, let's move on to the general concepts and procedures for designing and printing pages and covers.

Chapter Four
Page Design and Printing

1. Some Quick Notes on Typography

There are lots of places to learn about page design. There are any number of books and courses out there. Whatever page layout program you use will likely come with a booklet on the subject. Therefore I'll limit myself to a few quick notes on the very basic points.

Book printing uses real quotation marks " and " as well as proper em-dashes — and not the typewriter equivalents (" and a double dash --) used in a lot of word processing. Standard manuscript typing practice is to put two spaces after a period that ends a sentence. Standard typesetting practice is to use only one space. Follow manuscript standards while typing manuscripts and typesetting standards when setting type. There is usually some way to make these changes automatically via a feature built into the program, or by writing a macro yourself.

Don't go crazy with fonts and typefaces. Choose a nice, easy-to-read font for body text, another for headlines and subheads and titles and so forth, and maybe, at most, a third or forth for specialized text like captions or quotations. If you are going to use outside printing, make sure that whoever does the printing for you has physical and legal access to the fonts you want to use.

Keep the use of **bold** and *italic* to a workable minimum, and avoid the use of underlining in body text. As should be apparent from this paragraph, *italic* works pretty well in body text, but **bold** and underlined text just looks strange and tend to throw the reader off. As a general rule, use italic sparingly, bold very occasionally and only in certain types of layout, and underlining not at all.

In the vast majority of cases, you'll want body text with a serifed variable-pitch font, like the Times font used in this body text. "Serifs" are the strokes added to the basic shapes of a letter. "San-serif" fonts like Helvetica (used in the running heads of this book) do not have these strokes. These simpler letter-shapes can work better than serifed letters for headlines and short pieces of text, but are harder to read in long blocks of text.

The typeface that is best for one job can be a very bad choice for another task. In manuscript form, use a fixed-width font like Courier, with a ragged-right margin. In the wide margins of a typewritten page, Courier ragged right is far more readable than, say, Times, either ragged or justified. I know from experience that proof-reading a hundred-page manuscript in Times 10 point can get very hard on the eyes. Courier is also far easier to deal with for editing purposes, such as manuscript mark-up and word-count estimating.

Variable-pitch type is more readable in the narrower margins and right-justified paragraphs used in books and magazines. In your book pages, use fully justified text, with both the left and right margins of the text squared up and aligned.

In short, use the right tool for the job. Use Courier and ragged-right on your manuscripts, and you'll make your editor happier. But book pages should have variable-pitch fonts and right-justified margins.

Modern word processors and page-layout programs have made it too easy to create fancy, complicated, ornate, excessive layouts that get between the words and the reader. Keep things as simple as possible.

Design should aid understanding, not impede it. If putting a box around a block of text and setting it in a different point size will help set it off from the main text and make things clearer, that makes sense—because it literally helps the words make sense.

But if you've invented a way-cool font based on the Tolkeinesque Elvish language, and want to put it into your biography of your grandfather because you have it in your font folder, forget it. What do elves have to do with your grandfather? (And, by the way, that example is not pulled out of thin air. I knew a writer and artist who invented his own utterly indecipherable

Elvish font, and was determined to use it in his book. The book, so far as I am aware, has yet to see print. Draw your own conclusions.)

Add different layout elements and styles if you need them, and if they help the words. Don't invent fancy layout tricks and look for excuses to plop them down into the middle of a layout where they have no business being.

2. Choosing the Proper Software Tools

Broadly speaking, there are two kinds of computer program for dealing with words: word processors on the one hand, and page-layout or desktop publishing programs on the other. Which one is right for the book you are doing? No one program of any kind is best for every possible job. It's a question of the right tool for the right job, of horses for courses.

The names of the two kinds of text-manipulating program are enough to tell you what sort of task each emphasizes. Word processors are most concerned with words, with editorial content. Page layout and desktop publishing programs are more concerned with layout, with appearance, design, and style.

The two are not enemies or opposites (though I know of plenty of publishing operations where the editorial and design or art departments think in those terms.) The two work in concert. The best-written book in the world is of no use if it is printed in illegible print on bad paper. The best design work in the world is empty and pointless if it is used to support words that are mere meaningless drivel.

The two program categories have a lot of overlap. Lotus Word Pro or Microsoft Word or WordPerfect or any number of other word processors can be used to produce a pretty good, or even first-class, layout. Word processors can do lots of page layout jobs, such as running headers and footers, diagrams, drop-caps, tables and so on.

Contrariwise, PageMaker or Quark or InDesign or any of the other page-layout programs could, at least in theory, serve as a word processor. Many DTP packages include text editors, spell checkers and other writing tools. However, these programs would be pretty unwieldy for someone who just wanted to do some light typing.

No one tool is best for every job, and there is more than one kind of book-on-demand. I discovered how true that was in the course of writing and revising this book on book-on-demand writing.

I usually write novels, where I get to make everything up, where things don't change unless I want them to change, and where I don't need to keep things up to date. The book-on-demand projects I did before this guide have followed that pattern. Once it's done, it's done.

Doing a book like this guide, a book full of information that needs to be accurate and requires constant updating, where the text changes nearly every time I print it, and where I'm learning more about my fast-changing subject on a nearly daily basis, is a quite different job than doing a layout on a novel.

Once complete, the layout on a novel, or an historical text, is not going to change much, aside from the correction of the occasional late-found typo. You can do the layout once, by importing the text from a word processor into a page-layout program, and be done with it. In such a case it makes sense to keep the writing and design of the book largely separate from each other. The designing starts only after the writing is done. One job stops, and then the other starts. There's no point in letting the demands of one task distract you from the other.

But in a book that need updates, that's not the case. Page-layout programs focus on words as blocks of type, as design elements. They are therefore inherently more awkward for doing rewrites and inserts. It's quicker and easier to do your updates in a real word processor.

This does mean letting design quality take a slight back seat to content. PageMaker, for example, does a better job of kerning text, of hyphenating gracefully, and so on, than any word processor. A page done by PageMaker, especially a complex one, will often look better than the same page done in Word. But prior to this edition of the *Guide*, I was reluctant to give up the flexibility of a word processor for the more rigid, more precise, and harder-to-update layouts I'd be doing in PageMaker. Now I have made the jump, and it will mean a few more steps in the process when I do an update. But maintaining the text while controlling the layout in a word processor was getting unwieldy.

In the last version of this book, I had all the indexing, address lists, line illustrations, running heads, and everything else in one word processing file, and that file was written in DeScribe, a very good but very obscure word processor that hasn't seen an update since the company went out of business in the mid-nineties. Things were getting very creaky.

Now I have the names and addresses in a database, I have redone the illustrations in Illustrator, the words were imported from Microsoft Word, and PageMaker 7.0 handles the pagination, indexing, table-of-contents generation, and all the rest of it. Now I can let each program do what it does best.

FrameMaker, also made by Adobe, is supposed to be the best way to go for books that need constant updating, but I haven't spent the $600 I'd need to buy a copy and find out first-hand. The latest Adobe page layout program is InDesign, but I feel no need to spend *that* $600 or $800 either.

Not all page layout pages cost five or six hundred dollars. Microsoft Publisher and Serif's PagePlus go for under $100, and are often bundled into this or that package for free. These are less powerful programs, but you're not likely to need everything PageMaker or Quark can do.

Thus, among the first questions to consider when you start to lay out a book is, how often will you have to revise it, and how easy will that be with the software you are using? Will you have an index, appendices, a table of contents that will need repeated updates? How complex will your layout be in terms of illustrations, columns of text, and so on? Will a relatively simple and easy (or cheap) program really be able to do the job? Or will you need a more complex, expensive, and sophisticated set of tools?

Think ahead and choose a program that is going to produce

satisfactory results without making for extra work either on the first pass, or further down the road when it's time to rewrite.

3. Paper, Paper Grain, and Paper Curl

One frequently overlooked issue in regards to printing is the paper one prints on. There are several important characteristics you should consider.

Opacity

The paper you print on should be as opaque as possible, with very little see-through. If you can plainly and easily see the type on the opposite side of the page, your paper is not opaque enough. Paper opacity is measured against a standard scale. Most book paper ranges between 81 and 96 opacity. Evans suggests a paper with at least 92 opacity.

Paper Grain

The grain of the paper for both the book pages and the cover should run parallel to the book-spine edge. Paper with grain parallel to the long side of the sheet is "long-grain," while paper with the grain parallel to the short side is "short-grain." For one-up books printed in "portrait" mode, this means you'll usually want to print on long-grain paper. If you're doing two-up books printed in "landscape" mode, you'll want short-grain paper, as this will result in long-grain half-letter size pages. Most paper is long-grain. If you can't find short-grain letter-sized paper, take a stack of 11 by 17 inch ("tabloid size") long-grain paper and slice it in half down the middle, thus instantly converting it into two stacks of short-grain letter-size paper. If you're planning to do half-letter sized books, you'll be cutting that paper in half again, making it 8½ by 5½ long-grain paper.

However, here's fair warning on two subjects regarding short-grain paper. If you cut it yourself, make sure your cutting is *extremely* accurate, or else half your paper will be too narrow, and half too wide. Secondly, no matter who cuts it, be aware that paper curls with its grain, and paper curls when it is heated in a

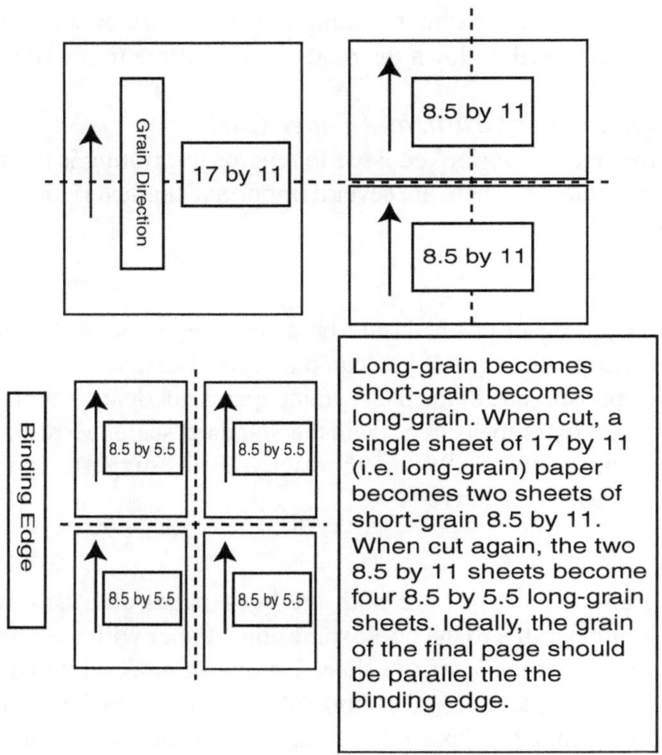
Grain Direction
17 by 11
8.5 by 11
8.5 by 11
Binding Edge
8.5 by 5.5
8.5 by 5.5
8.5 by 5.5
8.5 by 5.5
Long-grain becomes short-grain becomes long-grain. When cut, a single sheet of 17 by 11 (i.e. long-grain) paper becomes two sheets of short-grain 8.5 by 11. When cut again, the two 8.5 by 11 sheets become four 8.5 by 5.5 long-grain sheets. Ideally, the grain of the final page should be parallel the the binding edge.

laser printer. Taken together, this means that short-grain paper will curl side-to-side, rather than end-to-end. That can mean lots of paper jams. Be prepared to experiment, and to discover most papers will work just fine in your printer as long-grain and terribly as short-grain. Some printers are better than others at handling short-grain paper.

Here's another note that may clear up some confusion—or cause more. The normal convention in paper sizing is to give the paper size with the grain-side dimension listed first. Thus, the term "11 by 17" should mean the paper has the grain running parallel to the eleven-inch side, and is therefore short-grain. If it's called 17 by 11, that's supposed to mean the grain is parallel to the seventeen-inch side, and it's long-grain.

Unfortunately, this convention is almost always ignored when it comes to the sizes of paper used in book-on-demand. Just about all the paper you see called 11 by 17 should really be called 17 by 11, and nearly all the 8½ by 11 should really be called 11 by 8½.

As just about everyone from the paper manufacturers on down gets this notation wrong when it comes to letter, legal, and tabloid-size paper, I am not going to worry much about getting it right myself in this book, outside of this brief section. But the rule is worth knowing, because it is followed when it comes to larger paper sizes, (such as 12 by 18, and 13 by 19, and so on). And, if you do see paper listed as being 17 by 11, or 11 by 8½, that almost certainly means the maker is following this convention.

Paper Curl and Two-Up Printing

For many of the book projects a book-on-demand publisher might consider, it makes good sense to print multiple pages on one sheet of paper. (In this guide, "page" will mean the finished book page, while "sheet" will refer to the initial piece of paper on which one or more pages can be printed. Print two pages on each side of a sheet, and print on the front and back of that sheet, and you can get four pages on one sheet.) Printing two pages on one side of one sheet is called "two-up" printing.

Note that two-up requires printing in "landscape" mode, as opposed to "portrait" mode.

Doing two-up not only saves paper, but a great deal of time as well. Unless you are doing extremely complex pages with lots of graphics, most printers won't take any longer to print two pages per sheet instead of one. And, if you do two-up, you won't have to load the printer as often.

As should be obvious from the diagram, trimming the two-up pages to the final rough size requires one cut for every two pages, whereas one-up requires two cuts for every one page. While this sort of timesaving sounds trivial when considering one, or ten or twenty, sheets of paper, or even one copy of a book with two hundred pages, it rapidly becomes significant when you're doing ten or twenty or a hundred copies of that two-hundred page book.

Getting your pages into two-up format, with the pages in the right sort of logical sequence, can be a bit tricky, but it is certainly doable. Many page layout and word-processing programs include some facility for imposing pages properly. We'll discuss this under the topic of *Page Imposition* a bit further down.

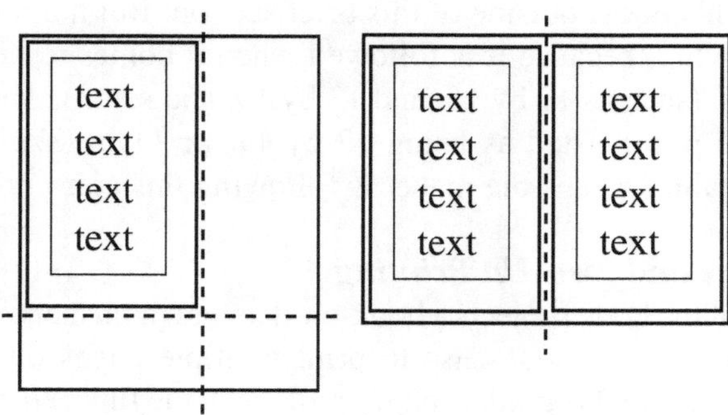

A page printed one-up on a portrait-oriented sheet, and two-up on a landscape-oriented sheet. Dotted lines indicate paper cuts.

Two-up pages end up with their curl at right angles to what it was in the sheets they were cut from. That can cause trouble when it comes time to assemble the pages into a book.

Most of the curl produced by paper run through a laser printer is caused by the printer's cylindrical fuser drum, which heats the fuser-facing side of the paper more than the other. This drives moisture out of the fuser-facing side, causing it to dry and shrink. According to one printer technician I spoke with, some additional curl is caused by moving the heated paper through the tightest curves in the printing path, known as the "pinch." This in effect bends the paper while it is hot, and the paper retains the bend as it cools. Dr. Evans disagrees with this theory.

One way to deal with curl would be to get a printer that can handle 17 by 11 paper and then load the printer with letter-sized short-grain (that is, 8½ by 11) paper. Load the paper sideways, feeding it with the eleven-inch wide side facing into the printer. That will cause the grain to run parallel to the paper path (meaning few or no jams and reduced curl), while at the same producing final pages that are half-letter sized, 8½ by 5½, that will be long-grain, with the grain parallel to the spine edge of the book.

According to the printer tech I talked to, simply using a tabloid-capable printer will reduce paper curl, even if you feed standard long-grain letter-size (i.e. 11 by 8½) paper with the 8½ inch side faced into the printer. The pinch on a tabloid-size printer is much wider in diameter, and thus the paper is wrapped far less tightly as it is heated, causing less curl.

All this talk about long-grain and short-grain might be confusing, so let's sum up the basic points.

- Paper has grain, just like wood. It bends, cuts and folds more smoothly parallel to the grain.

- Paper feeds best through a laser printer when the grain is parallel to the printer paper feed path.

- Virtually all letter-sized paper is long-grain, and should more properly be called 11 by 8½, rather than 8½ by 11. Nearly all tabloid-sized paper is long-grain, and should be called 17 by 11, rather than 11 by 17.

- Cut 17 by 11 long-grain paper in half across the long axis, and you will produce short-grain 8½ by 11 paper. Cut this stock in half, and you will produce long-grain 8½ by 5½ paper.

- For best results, paper grain should be parallel, and not perpendicular, to the spine, or binding edge, of the final book pages. Either grain direction will work, but book pages with the grain perpendicular to the spine are a bit stiff and not as easy to handle. The ideal is therefore to print with grain parallel to the paper path, and then bind with the grain parallel to the spine. If you're lucky, your printer will be able to handle short-grain paper without choking. I have two laser printers. One can handle short-grain and the other can't. You might find that you have no choice but to use long-grain paper. If so, you'll still want to get the grain of the final page parallel to the binding edge

Here are four possible ways to accomplish that goal:

1. Use full-size pages, 8½ by 11 inches, allowing room for final trim. Use standard long-grain paper.

2. Print smaller pages one-up in "portrait" mode on standard long-grain letter-sized paper, and feed the sheets into the printer with the narrow edge of the paper faced into the printer. Bind along the long edge of paper. Cut down pages to final page size, thus discarding a large part of each sheet of paper. This might be wasteful, but it works.

3. Print two-up in "landscape" mode on short-grain paper, and feed pages with the wide edge of the paper faced into the printer. (This requires a printer that will accept wider stock—in other words, a tabloid printer.) Cut the sheets apart, thus producing two stacks of pages on long-grain 8½ by 5½ paper. Bind along the 8½ inch edge.

4. Print pages one-up on long-grain paper that has already been cut down to the final page size, for example 8½ by 5½, or 9 by 6, or 10 by 7. This will require a printer that can handle customer paper sizes. Bind on long side of paper.

I should note that the above discussion all assumes that your final book will be bound in "portrait" mode, with the long axis of the book vertical as you hold it for reading. If you're doing books that will be bound in "landscape" mode with the long axis horizontal, you quite literally need to do things the other way around. How you manage it will depend on your page size, but just keep two main points in mind as you work it out.

- Your printer won't jam as much (or at all) if you keep the paper grain parallel to the print path.
- Your books will open more easily, and be easier to handle, if the paper grain is parallel to the spine.

And that is about as much as I care to discuss paper grain. Let's go on to talk about something more exciting —like how many pages you can fit on one sheet of paper.

4. One-Up vs. Two-Up Printing

Despite the advantages of two-up printing, the average book-on-demand printer can't do it for any book over 8½ inches high or 7 inches wide, that being half the size of a sheet of legal-size paper. Most laser printers will not accept paper larger than that.

Anything over 8½ by 7 inches wide size must be printed one-up. Many books are printed to the size of 6 by 9 inches, or 7 by 10 inches, neither of which can be printed two-up on letter-size or legal-size paper, and therefore must be printed one-up. This means throwing away a large fraction of each sheet of paper, wasting material and money. You might as well print at full letter size, 8½ by 11 inches. Many book-on-demand printers prefer half-letter or half-legal size book pages, that allow two-up printing, so as to avoid wasting so much paper.

However, as we have seen already, and will see again later in this chapter, two-up printing does create a lot of work and confusion when it comes time to get the pages in order for printing. One-up pages automatically fall into order. For some workers, and some projects, avoiding page-imposition, which we'll discuss later in the chapter, might be worth wasting a lot of paper.

Throw scrap into the recycling bin to avoid feeling too guilty about waste.

5. Twin Two-Up Printing

This is an extremely useful variant on two-up. Basically, it is two-up with the right side being a "twin" of the left side. The first sheet has two copies of page one, the second has two copies of page two, and so on. You are actually printing two copies of the book at the same time—one copy on the left, and one on the right.

Under certain circumstances, you might run into trouble with pagination in twin two-up. For example, some word-processors might get confused and want to auto-paginate the pages such that the second copy, that is to say right-hand pages, would be numbered one higher than the left-hand pages. However, the various forms of imposition software, discussed below, simply move around, re-scale, resize, and copy page images, picking up the pagination from the original, one-to-a-sheet, pages. This eliminates the problem.

There are two advantages to two-up twin. First, it avoids the muddle of stacking pages incorrectly. With normal two-up printing, the pages can be printed in at least two different sequences, and it is easy to arrange the final page-stack incorrectly. (See the discussion of page imposition below.) With twin two-up, the left and right stacks are independent of each other, and automatically come out in the right order. The second reason has to do with paper curl. When paper comes out of a laser printer, it has a curl. The curl may not be noticeable with one sheet, or with a short stack of sheets, but with a thick stack of paper in humid weather, the curl produced by some printers goes from noticeable right on up to out-of-control. With normal cut-and-stack two-up printing, with, say, pages 1-50 on the left side of the sheet and 51-100 on the right side, you'd get half the book pages curl one way, while the other half curl in the other. This looks strange, and can weaken the binding. All the pages of a twin two-up printed book curl the same way, eliminating the problem.

Curl needn't be forever. Stacking paper under weight will reduce curl, but it can take a long time for a laser-printed page to go completely flat. I've found curl seems to flatten out faster and better if all the pages are curled in the same direction—another argument for twin two-up.

6. Duplex Printing

"Duplex" simply means "double-sided," the ability to print on both sides of one sheet of paper. A printer that can automatically do duplex printing is called a duplex printer. (A printer that can only print one side is sometimes called a "simplex" printer.)

Duplex can refer not only to the printer, but to the act of printing. Even a simplex printer can do duplex printing. All you have to do is print the odd pages on one side of a stack of paper sheets, turn the stack over, and print the even pages on the other. Nearly every word processing and page-layout program has some provision to do odd-only and even-only printing.

However, it is not quite that easy. Suppose you have a hundred-page book, and run off odd pages 1-99 without trouble, then flip over and print even pages 2-100 on the other side—and get a paper jam, or two sheets feeding together, or some other common printing problem, on page 2. From page 4 to 100, the odd pages will be one off from the even pages on the other side of the sheet.

To make matters worse, as we have discussed more than once already, laser printers cause paper to curl. The printer uses heat to bond the toner to the paper, and that heat drives water out of the paper, warping it like a board of wood drying after it's been left out in the rain. Obviously, curled paper does not feed into the printer as well as flat paper. This means the second pass through the printer is much more hazardous for the paper. Your odds of a paper-jam far more than double when you send the paper through twice.

Paper curl problems don't stop when the printing is over. Trying to bind a stack of excessively curled paper can get to be a real struggle. As we have seen, books printed two-up with the two halves then stacked together can make half the paper curl one way, and half the paper curl the other. This does not look good, and can weaken the binding. I don't know of any single solution to this problem that will solve it in all times in all places. There are ways to get around it nearly all of the time in most circumstances.

- Store your paper flat in its original packaging until you use it.
- Control humidity. If the humidity in the room where you store paper oscillates, the center of a stack of paper will have less chance to absorb and release water than the top and edges, causing great variability in water retention, and thus, ultimately, curling. If you want to go all-out, consider storing the paper in a big Tupperware container with a few of those desiccant packets.
- After printing the first pass, roll the paper against the curl. This will sometimes eliminate a fair amount of the curl.
- After printing the first pass, stack the paper under weight for a while before doing the second pass.

Note that, because ink jet printers and impact printers use little or no heat in order to put ink on the page, paper curl is far less of a problem, and you will almost certainly be able to print the second pass without taking any steps against paper curl. However, ink jet printers are not a perfect solution. They rarely, if ever, put down as sharp an image as a good laser printer, generally print far too slowly to be practical for production book page printing, and their operating costs (i.e., cost of ink cartridges) tend to be much higher on a per-page basis. The most serious problem with ink-jet printed pages is that most ink-jet inks are water soluble. Sweaty hands will smear the pages.

Two good ways to get double sided printing are:

1. The purchase of a laser printer with a built-in duplexer, or one that can take a duplexer as an add-on accessory. These machines are designed to deal with paper curl, and do so, if not perfectly, at least extremely well. A duplexing printer prints one side, runs the sheet through a special paper path that turns it over, and then prints it on the other side. Since both sides are printed in one pass through the printer, the problem of getting odd and even pages out of sequence is eliminated.

2. Print your pages out on a simplex printer, then go to a local copy shop and have them copy your single-sided sheets onto double-sided sheets. Obviously, it pays to be careful and explicit in explaining to the clerk exactly what you want. You can often get a better deal on printing by agreeing to collect it a day or a week later, rather than having it done while you wait. It might well be worth while to phone several shops, explain exactly what you need, and get price quotes.

The down-side of copying your pages this way is that it will cost more per page than printing it yourself, and you will lose a certain amount of print quality. It will, however, eliminate the headache of manual duplexing, or the high cost of a duplexing printer.

7. Page Imposition

"Imposing" a page means to position and orient it on a sheet in preparation for printing. The term and concept are borrowed directly from conventional printing. Nearly all conventionally-printed book pages are printed, not one-up or two-up, but four-up, sixteen-up, or even thirty-two-up. An imposition might print eight pages on one sheet—four on the front, and four on the back. Once both sides are printed, the sheet is folded into a signature. Some impositions require pages to be printed upside-down relative to each other in order for the page to be oriented properly once the large sheet is folded and trimmed into a signature. Once the folding is done, any edges with folds, except for the interior margins, are trimmed off so that pages can open freely. The result is a signature of eight or sixteen or thirty-two pages.

In the case of most high-quality hardcover books, each signature is sewn together into a unit with thread run through the spine-edge fold. The various signatures are then assembled in the proper order and sewn or glued together along the spine edge.

In the case of most paperback books, the folded signatures are not sewn, but simply assembled in order. The binding machine clamps onto the signatures. A special cutting head grinds, or *mills*, the spine edge of the signature, exposing the inner edges of all the pages, so that each book page become a separate piece of paper. With the binding edge of each page exposed, the binding glue is applied, and the pages are held together.

The situation is a bit less complex for the average book-on-demand printer, who only needs, for the most part, to worry about two-up impositions, but the concepts are the same. The book pages need to be positioned properly on the sheets of paper for printing, and the pages need to be placed in the right order for printing.

Consider a very simple example, as seen in the diagram below. In order to make the page order come out, a four-page booklet is printed on one sheet of paper, with page 4 on the left of the exterior side, page 1 on the exterior right, page 2 on the interior left, and page 3 on the interior right.

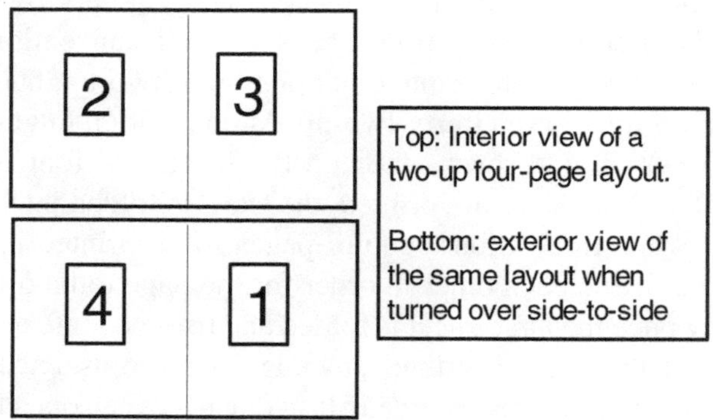

Top: Interior view of a two-up four-page layout.

Bottom: exterior view of the same layout when turned over side-to-side

For an eight-page imposition on two sheets of paper, the proper order would be:

- sheet one: exterior left, page 8; exterior right, page 1; interior left page 2, interior right, page 7.
- sheet two (nested inside sheet one): exterior left, page 6; exterior right, page 3; interior left, page 4, interior left, page 5.

I did that in my head. I think it's right. But the point is that figuring such things out is a nuisance, even for a very simple example. What about a real-life 323 page book? How do we do the page ordering?

Before we go any further with this subject, do take note of a few key points:

- If you do a four-page-to-a-sheet imposition, you're going to have to have a final number of pages divisible by four. If you're doing X pages to a sheet, your final number of pages must be divisible by X, and extra pages must often be added to make that happen. If you've ever wondered about the two or three utterly blank pages at the end of a book, they are there to make the signature come out even. (Because the pages are cut apart in paperback books, these blanks are often discarded before binding. Some publishers fill up blank end-pages with advertising.)

- Because they are based on folding paper in half, thus doubling the number of pages with each fold, the basic impositions work in halves and doubles, in the sequence 2, 4, 8, 16, 32, 64. No one does impositions of eleven pages to the sheet.

- Simple double-sided printing, with one complete page on each side, can be thought of as the simplest possible page imposition, the only one in which the print-order and the page-order are identical.

To sum up, imposition is complicated. Fortunately, it's a repetitive job that can be done following rigid rules: in short, it's the sort of thing computers are good at. Some, though by no means all, word-processing and page-layout programs include booklet-making features that calculate simple impositions. Some of these programs manage this task better than other.

Adobe's PageMaker page layout program has a booklet maker (available in various guises in version 5.0 and up). It does a pretty fair job, but it can get confused, and can take forever, on longer jobs. It works better and faster with more memory and more available drive space. It works far better if the entire book is in one file. At least in some versions of the program, if the book to be imposed is in two or more PageMaker files, the formatting of paragraphs that span from one page to the next will be thrown off.

Microsoft's Word for Windows doesn't seem to have such a capacity that I can track down, though there are third-party macros for it that do the job. WordPerfect has had booklet printing since 1985. Lotus WordPro includes a booklet-printing feature. I don't know much about how they do this sort of thing on Macintosh.

For users of Windows or Macintosh programs with no built-in booklet-printing feature, the excellent program *Clickbook*, from *Blue Squirrel*, can ride to the rescue. It will intercept the output usually sent to your printer and scale and rearrange the pages to fit into two-up format, or into twin two-up, or into a number of other formats.

Fineprint, a printer driver from *Fineprint Software* performs similar functions. It can do a few tricks Clickbook can't, and vice versa. Try them both and see which works for you.

Just by the way, I have found that a document laid out for a legal-sized page (8½ by 14) with somewhat oversized type, say 16 or 18 point, scales and rotates very nicely into a 5½ by 8½ page with about 12 point type, when processed by Clickbook.

If you have a simplex printer, Clickbook will also do a nice job of walking you through the step-by-step process of printing your book double-sided. I could go on, but suffice to say that Clickbook can perform lots of other neat tricks, and that it costs about $50. It will make your life at least $50 easier, and is well worth getting. Fineprint goes for about $40, and while I have only experimented with it briefly, my first impression is that it's pretty nifty.

A word or two of caution concerning Clickbook, Fineprint, or any similar program: test it with your word-processor or DTP program *before* proceeding with a complex layout. Clickbook works fine, but the interaction between it and certain programs can be a little complicated.

To oversimplify a bit, under certain circumstances, when some programs use Clickbook for output of a half-letter page, Clickbook needs to have the half-letter page centered on a full letter-sized page image on the screen. With other programs, it needs to have that half-letter size page in the upper left-hand corner of the full-size page image. If you prepare a layout for one eventuality, and come across the other, you'll create a huge amount of work for yourself, and have to redo the entire layout.

As the various WP and DTP programs get upgraded, and as Clickbook itself evolves, what works with one combination of programs likely won't work exactly the same way after the next upgrade. Do some tests before you start in on production work.

That brings us to a good general rule. Test the tools you're going to use before you need to use them. Have the software you plan to use to produce two-up layouts in hand, and experiment with it, before you start laying out your book. Understand your software well before you use it, or you will invent a lot of work for yourself.

With or without Clickbook or Fineprint, give careful thought to how you are going to do page imposition before you start with your layout. Do a few test runs, with a document comparable in length to what you plan to print. (Lots of the built-in booklet making programs and macros and so on can choke up or crash altogether on big print files.)

As I have already noted, the further along you are in a given project, the tougher it is to fix mistakes. Many of the booklet-making systems create a new document with the text chopped up and the pages reordered. Fixing a typo (and especially one that changes line-count) in such a file can wreak havoc, if it's possible to fix it at all. In such cases, it's probably easier to go back and fix the original layout document, and then re-impose it.

It's a good rule of thumb to play around with all the equipment, software, and procedures you'll need for doing book-on-demand before you set out to do actual work. This goes double for page imposition. Give yourself a chance to discover nasty surprises early on. Deadline time is no time to discover that your imposing software crashes if you use Helvetica 13 point italic.

8. Using Custom Paper Sizes To Avoid Imposition and Cutting

Obviously, imposing can be a bear. One way to avoid the headaches of dealing with it is to print everything out one-up, cut away the blank parts of the page, and be done with it. Aside from being ungainly, sloppy, and downright inelegant, such a solution is almost criminally wasteful. You'll throw away as much paper as you use.

However, you can avoid the need to struggle with page imposition and also avoid massive paper waste if you have a printer that can handle custom paper sizes, as discussed in the printer section of the previous chapter. If your printer has adjustable input trays, you could cut your stock down (or have it cut down for you) to the final page size (of, say, 8½ by 5½ or 8½ by 7 or 6 by 9) and print on that size paper. With just a little bit of planning, you could see to it these cut-down sheets were long-grain (see diagram on page 70), which would make for book pages that handled better in the book, and would also eliminate a lot of paper jams. (I use custom-sized paper for a number of books I print. I order 9½ by 6¼ paper. That allows for a quarter-inch trim from the top, bottom, and spine-opposite side of the paper, and gives me a standard 9 by 6 inch book.)

You might even be able to use this trick to avoid the purchase of a guillotine cutter. You should be able to find a paper supplier who will cut paper to a custom size for a flat fee per order. If you order enough paper at a time, the cutting charge won't be anything much. Print on these pre-cut sheets, and you won't need to cut the pages apart. If you don't have a guillotine, you won't be able to do a final trim on the assembled book, but if you work carefully and neatly, you *might* be able to get away without that step.

9. PostScript, Acrobat, and Page Imposition

This chapter section deals with a tremendously complex subject—or rather, a tremendously complex set of related subjects—that we'll to try and cover in as little space as possible. Briefly put, *PostScript* is a programming language, used primarily for page layout, though it can perform many other functions. It was until recently the *de facto* standard for sending and receiving page-layouts to outside print shops. While still much in use, it is being edged out by its near relation, *Acrobat*. Both Acrobat and PostScript are owned by Adobe. PostScript is a high-end feature on printers for the home office, but totally standard in any size print shop.

When you print to a PostScript printer, what you (or rather your computer program and you) are actually doing is writing a PostScript program, which is then sent to another computer, inside the printer—a computer that understands the PostScript programming language. The printer-computer then runs the program, which produces output in the form of printed pages. That same PostScript program can be saved to disk (and edited by hand, to a certain degree) and then sent via disk or network connection to any other PostScript printer anywhere in the world. Within certain limits, and with one or two caveats, a standard PostScript file will print properly on a 300 DPI laser printer or a 2400 DPI linotype machine, but the better the resolution of the printer, the better the printout will look.

PostScript files can also be manipulated by other software. *GhostScript* is a program available for Windows, Macintosh, Unix, Linux or OS/2. It will display the output of a PostScript file on screen, and is extremely useful for examining and editing PostScript files without using up paper. It will also print a PostScript file to many non-PostScript printers, though not at any great speed. The *GhostView* "front-end" program, available for Windows and OS/2, makes dealing with GhostScript much easier. Fair warning: neither Ghostview or Ghostscript are bug-free.

GhostScript and GhostView are available at no charge to most users, and can be downloaded off the Internet, though there are slightly complicated copyright restrictions for some forms of commercial use. See under "GhostScript" in the *Names and Numbers* list at the back of this book to learn where to get these programs. And, of course, there are many other PostScript utilities out there.

Specialized page imposition programs can take a PostScript file, reorder, resize, and reorient the pages and perform all sorts of other tricks. A file so modified can likewise be saved to disk and/or sent to any PostScript printer (again, within various practical limits).

One advantage of PostScript is that, once the final print file is completed, the program that created it (PageMaker, WordPerfect, etc.) is no longer needed to print further copies. Lots of things can go wrong, but in theory you can take your PostScript file to any PostScript printer in the world and print out your file all by itself, merely by sending a copy to the printer.

As far as getting that file to the printer goes, there are simple PostScript downloaders available for the Macintosh computers, while the easiest thing to do in the DOS-Windows-OS/2 world is to get to a C:> prompt and use the copy command. For example, COPY FILENAME.PS LPT1 would send filename.ps to the printer on parallel port 1 (LPT1). PostScript purists frown at this sort of "one-way" control of the PostScript printer, as it does not allow the printer to send back results information or be controlled interactively. All that is true, and you can screw things up royally with the copy command—but the trick will work, assuming your PostScript file is properly formatted.

Adobe *Acrobat* creates and displays PDF (Portable Document Format) files. The PDF format is based in large part on PostScript. Acrobat is gaining increased acceptance as a standard file format in the printing industry. Well-written PostScript files are said to be "portable" because they can be read by any PostScript printer, imposition program, or device. In similar vein, files in the PDF format are likewise said to be portable: they can be viewed by any computer, of any type, equipped with *Acrobat Reader* or other suitable viewer.

Adobe gives Acrobat Reader away for free in order to build the market for the full Acrobat program, and for the *Acrobat Distiller.* The full Acrobat program sells for about $250. The Distiller also ships with PageMaker. Distiller can create PDF files from PostScript print-to-disk files. The full Acrobat program allows you to edit and manipulate PDF files that already exist. Acrobat Reader will show PDF files on screen, and can print them to paper. There is also a Windows printer driver that can create PDF files, but supposedly the files it makes are not as slick or high-quality as the ones produced by Distiller or the full Acrobat program.

GhostScript, described above, can read and create PDF files. It is becoming more and more common for various other programs to generate their own PDF files. Fineprint Software make pdfFactory, which sells it two versions for about $50 to $100. There are now a number of sites on the World Wide Web that will take your file, in various formats, and generate a PDF for you. Some of these sites are free, some are not, and features vary.

Bear in mind that the quality of PDF files can vary a great deal, as can the file size. For example, because even a low-quality bitmap image can take up a lot of room in a file. Let me indulge in a bit of techno-gibberish to explain. It can be easier to generate bitmaps of text during PDF creation instead of handling fonts, it is often the case that a truly huge PDF file can look bad, while a smaller PDF of the same page, generated with the text treated as text, can look a lot better.

The main advantage of Acrobat, insofar as book-on-demand is concerned, is that the files are portable and at the same time insure that what-you-see-is-what-you-get.

An Acrobat file will look the same on an low-end IBM clone running Windows 3.1 or on a souped-up Macintosh, or on a Unix machine. Print a PostScript file to disk from any program, use Distiller or Acrobat to create the PDF from that, and then you can send the file to a printshop, or the sort of Publishing Service Provider discussed in Chapter Nine, and be confident the printers are seeing, and printing, what you are seeing. Assuming everyone you send the file has a free copy of Acrobat Reader, they can all see your Ventura or PageMaker or Word pages, even if they don't have those programs. Everyone can actually see the pages before they print, and know what's supposed to be there.

Acrobat will print files to pretty much any printer you can get a driver for. Obviously, here again, the better your printer, the better-looking the print-out. You might get it to work, but a file printed to a nine-pin dot matrix printer won't look all that great.

There are positives and negatives to using Acrobat, and to using PostScript.

Acrobat Advantages

- Acrobat Reader presents any Acrobat file in a fully what-you-see-is-what-you get form. (At least in theory. There are always bugs, glitches, and exceptions.) It is easy to read one page or print one page, and easy to navigate through a file. PostScript files are much tougher to page through, absent specialized software. Sloppily written PostScript can produce files that are total rat's nests, and are all but impossible to navigate.

- It is possible to print single pages, or a group of pages, of a PDF file. PostScript files are basically an all-or-nothing deal. Without using specialized software (like GhostScript) you can't print just part of a PostScript file. Version 3.0 of Acrobat Reader did not allow for printing out odd pages in one pass and even pages in a second pass (something you'd need to do if you had a simplex printer and wanted duplex output.) Various third-party programs can make this happen for version 3.0. Version 4.0 and up provide for odd-even printing.

PostScript Advantages

- PostScript files are ASCII text that can be edited by just about any word processor, though they can get to be pretty indecipherable at times. They are, after all, computer programs. It is, however, possible to track down and fix a typo in a PostScript file. Acrobat files aren't even that readable. They can only be edited and manipulated by specialized software.

- PostScript files can be sent directly to a PostScript printer by themselves with no outside software. Acrobat files must be printed by an Acrobat-aware program, like Acrobat Reader.
- PostScript files can be produced by any program that prints to a PostScript printer. Acrobat files must be produced by a program like Adobe Acrobat Distiller, or the full Adobe Acrobat program, the Acrobat Writer printer driver, or a third-party program that, to one degree or another, emulates one of those programs.

Another factor is the question of file size. Below is a list of file sizes in the various formats for a previous (and much shorter) version of this book:

File Format	Size
DeScribe Word Processor file	638,629
PostScript Print-to-Disk File of above generated by DeScribe (two-up pages)	1,114,241
PDF File generated from above PS file	339,532
PostScript Print-to-Disk File generated by Acrobat from above PDF file	731,183

All of these files, when sent to the printer, produced virtually identical results, in spite of the vast range in file size. The greys on some of the graphics varied a bit, but that was the only noticeable difference. PS code can vary a lot, and there are lots of different ways of coding the same final marks on paper. By using different drivers and techniques, I have produced PS files for the same book ranging from 600,000 to 3,500,000 bytes! In other words, some PostScript printer drivers produce very bloated code.

The relative sizes of the two PostScript files above demonstrate another use for Acrobat. It can serve as a PostScript "distiller," producing tighter, faster-running code.

One important point: graphics take up a lot of space in any file format. A graphic-intensive Acrobat file could take several hundred thousand bytes for a document that was only a few pages long.

But back to page imposition. Most page imposition programs start from PostScript code, and are intended for the professional print shop. They are overkill, and overpriced, as far as a home book-on-demand printing operations is concerned.

If you search on the World Wide Web using the terms "Postscript" and "Imposition," you'll find a number of PostScript imposition packages. They start at well over $1,000, and thus not of interest to most readers of this book.

Some imposition programs can deal with PDF files. An Australian firm, *A Round Table Solution (ARTS),* offers a range of PDF-manipulation programs. *PDF Snake* is a (relatively) inexpensive PDF page imposition tool, costing $240. A British company, *Quite Software*, sells a program called Quite Imposing, that does page imposition for Acrobat files. It sells for £199, or in a "professional" version for £389. Both companies offer demo versions that can be downloaded off their web pages. Quite Imposing requires the full Acrobat program. These and many other PDF manipulation programs are available at the website www.epublishstore.com.

There is one other more subtle advantage to using PostScript or Acrobat: when it comes time to print, they have a much smaller "footprint," meaning they use up a lot less of your computer's resources than a typical page-layout program or word processor. On my machine, at least, Acrobat Reader loads in about five seconds. PageMaker can take over a minute—and it gobbles up a lot of system resources that Acrobat Reader does not.

In short, PostScript and Acrobat can make things more complicated and more efficient, all at once. If you always print directly from word processor or page layout program, you may never need to worry about them. But if you expect to have someone on the outside print your text, now or later, or even if you just want to be able to print your finished work more easily, look into PostScript and Acrobat.

10. Adding Sheets From Other Sources

Nowhere is it written that everything in a book needs to printed in the same place at the same time. Assuming you're careful and don't get the page order messed up, there is no reason in the world you can't add, for example, a page or two of ink-jet color into your book, or a photograph, or whatever. You can drop in pages that have been printed by an outside shop. Or if PageMaker is kicking up a fuss about importing the graphic that CorelDraw can handle without any problem, well then, print that page in CorelDraw and tip it into the finished stack of book pages printed in PageMaker. Do bear in mind that some coated papers, such as some of those used for photo-quality ink-jet printing, may not bind well with some glues. Also be aware that ink from an image could rub off onto the facing page, leaving a faint "ghost" image. You can usually avoid these problems by using high-quality paper, or even by using a spray-on laquer or laminate, or by getting *very* fancy and putting in a sheet of tissue paper between the picture and the facing page. But that is just a bit much.

That concludes our whirlwind tour of the main issues involved in producing pages for book-on-demand printing. Now that we have the pages, let's take a look at what the pages will go in: the book cover.

Chapter Five
Printing Book Covers

1. Cover Design

Before we deal with the mechanics of printing covers, let's take a moment to discuss what will be on that cover, and how it will look, and what physical characteristics it will have. The following discussion assumes that you are trying to sell your book in a retail environment, where appearances count for a lot.

The one most important fact to remember is that your book cover is an advertisement for the interior contents. It should make a member of your target audience want to pick the book up and have a look at it. Don't just make your cover attractive. Make it attractive *to its target audience*.

If the author is an important draw, put his or her name in big, prominent type. If the subject or the title is likely to sell more books, play up that element.

If you have a series of similar books, make them similar enough in appearance that someone who owns the first will spot the second and third as part of the set. The books should not be identical, but should have a close family resemblance. They might, for example, each have the same basic layout, but with each having the title in a different color.

Keep it simple, and keep it legible. This is not the place for me to tell horror stories about books with covers that used designs and fonts that were so darned busy and interesting and exciting and different that no one could pick out the book title or the author. Let's just say I have reason to know it happens.

Take the time to show your cover design to people who don't know the title or the author's name. Make sure they can read that vital information off the cover quickly and easily from a few feet away, browsing distance in a bookstore. That advice might sound silly, but take my word for it, there are giant publishing houses that would sell more books if they followed such a policy.

Try and get some useful blurbs (that is, positive quotes from people who might know something about the subject or genre) on the cover. Quotes from reviews are also good to have. Work up clear, concise back-cover copy that explains the book.

2. ISBNs and Barcodes

Any book sold retail will need an *International Standard Book Number,* or ISBN, and really ought to have a readable barcode on the cover. There are lots and lots of programs out there—more than can be listed in this space—to do barcoding. Some of them are shareware, and some are commercially available. Basically, they will all generate a barcode from information keyed in by the user. There are also barcode fonts available, with each digit represented by its barcode equivalent. A company called *SNX* will email you a graphic computer file, an EPS (Encapsulated PostScript) file of your barcode if you send them the data that should appear in the code. Price, $15.00 a barcode. *SNX* also sell programs that allow you to produce your own ISBN barcodes starting at about $280. I have used the SNX service and software and been well-pleased with the results. *Barcode Producer* from *Intelli Innovations* is available for about $90 and has versions for Mac and Windows. I have not tested their products.. There are lots of other sources for barcoding software and service.

Lightning Source now provides free cover-layout templates for their customers via email. These templates include a barcode in EPS format. This book uses a barcode generated via this service.

The information contained in book barcodes consists of the arbitrary digits 978, the ISBN (digits only—the hyphens are dropped) an automatically generated and optional check digit, and, optionally, a five-digit field for the price or other information. I have seen the variants of this barcode format called ISBN, EAN13, EAN13+5, BOOKLAND, or EAN BOOKLAND.

The *R. R. Bowker* company administers ISBNs in the U.S. Their website, www.bowker.com, is a good place to start searching for further information on barcode specifications. Bowker sent me a list of barcode providers along with my ISBN kit .

There are lots of programs out there intended to do barcodes for grocery stores and the like. They will only print physical labels meant to be stuck on packages and so on. You need a program that will create a barcode *image* that can be imported or placed into your cover layout and printed out as part of the cover. In other words, you need a program that generates a graphical barcode image that can either be saved in a standard graphics format, or else electronically cut-and-pasted onto the cover layout.

Do not rescale or resize your barcode image. Use it in the size and shape you receive it, either from software on your desk or from a service. (It's all right to crop excess white space from around the barcode image. Just don't mess with the image itself, and leave *some* border around it.) Rescaling or resizing amounts to distorting the barcode—which means you've just made it a lot harder for the barcode wands of the world to scan your book.

The main thing is that what shows up on the final book cover should be a sharp, clear, readable barcode that will work. When I asked one independent bookseller what he thought of the covers on small press books, the first thing he complained about was barcodes that were too small for his code reader to read. Test your barcodes. Make sure the price, ISBN, and barcode are easy to find and clearly legible.

This means that if you are going to write barcodes, you should also have some sort of barcode reader in order to test the barcode in-house, before fifty or five thousand copies of your book go out with a wrong or illegible barcode. (And I should know what a bad goof that is—I've sent just such books out into the world. Mistakes cost time and money.) There are any number of wands and guns and scanners available on the market that will plug into your existing computer and do the job for you. If you can't find a cheap enough barcode reader, wander into your local bookstore and see if *their* barcode readers can read your barcodes.

3. Using a Commercial Printer

As is no doubt becoming apparent, a real commercial-looking

book cover is a complicated thing. It may be beyond the capabilities of some home publishers. There is no reason it couldn't be done with a really good color ink jet or color laser printer, but it might well make more sense to go to the outside for covers.

Figure out exactly what you want, and be prepared for some shopping and haggling when it comes time to buy covers. You want a printer who can go from your final design on disk and print directly from that. These days, most print shops will want to work from an Acrobat PDF format, but they'll want to make sure you've created that PDF as per their specs as to font-embedding, image compression, etc. Pick a printer, find out the specs he or she needs, and *then* create your PDF. Similar rules apply with printers who want Quark, PageMaker, InDesign or similar program files, or PostScript print-to-disk files. Many will want you to send your file over the Internet via email. Make it very clear you are doing covers that will need lamination or a protective coating.

But even if you do outside printing of your covers, you'll want to be able to run off your own in-house proof copy, just to be on the safe side and check on various details. It would be worth having a color ink-jet printer just for doing these cover proofs. Also insist on a proof of the cover—or you might end up with distorted and misplaced barcodes, (the way I did), or some other nasty problem.

One drawback to having outsiders do your covers is that they will likely want to print a lot of them, a lot more than you're likely to need at any one time. On the other hand, your cover is not likely to need revision as often as your interior pages, and you don't have to print books and covers at the same time. You could, for example, have 500 copies of your cover printed, but only use fifty of them on your first press run.

Another possibility is to get generic color covers printed up, with blank areas on the spine and front and back covers where you could drop in the title and art and cover copy using your own color or non-color printer. I believe the original Penguin Books covers did exactly this. The colored portions of the covers were printed in one huge batch, and then the titles and authors were printed in later. Obviously doing things this way would require thoughtful design, but it could give you a nice durable professional cover, and could give your books a shared graphic identity. If you do this, remember that all the ink should be on the page before you put down any laminate or varnish.

There are various digital color presses out there. These machines are designed to print full-color pages fast. They require little or no setup required, so turnaround times are reduced, and short press runs (of a hundred or five hundred) become practical—if still a bit pricey. You could easily pay $2.00 or $3.00 per cover done on one of these machines. Print shops often charge by the impression, and you can probably work a way to get two covers onto one sheet, which should help some. There are lots of color printing services out there, using everything from color photocopying on up to offset. Shop around.

However it is done, when it comes to printing, speed costs. Getting covers done overnight will cost a lot more than getting them done in a week. Plan ahead.

One little bit of decoding that might help. Printing rate sheets list prices for 4/0, 4/1, 2/1, and so on. These numbers refer to the number of colors printed per side. Four-color is full color, and black is just one color, so 4/1 printing would mean full color on one side, and one color (usually black) on the other.

4. Printing Your Own Covers

All of what I've said about covers that you have printed by a print shop still goes for covers printed in the home. Here are a few additional notes that apply mainly to home-made covers.

First, be aware that the heat from a office thermal binder, as discussed in Chapter Six, is enough to soften or melt laser toner. (Because they don't heat the exterior of the book significantly, the larger binders discussed in Chapter Seven and Eight won't cause this problem.) If you run laser-printed covers through an office thermal binder, the odds are pretty good that the heat of binding will make your spine letter blur or smear. (Just to be clear: the printing on the inside pages will be safe. That printing doesn't get hot. It's just the spine printing on the cover that is at risk.)

A material called release paper can help protect that spine lettering, at least somewhat. The basic job of release paper is to refuse to stick to anything. It's slick and waxy to the feel. It is essentially the same stuff as the backing paper on adhesive labels (and a few sheets of that backing paper might be all the release paper you ever need.) The real stuff is usually sold in photography stores. It is used in thermal dry mounting. Release paper placed between the cover and the binding machine's heating plate will offer some resistance to melting, but it will not offer perfect protection.

The ink-jet inks I have experimented with are not affected by heat, but moisture will definitely effect them. Under certain circumstances, water-based glues could wet your cover stock excessively and mar the appearance of the ink-jet ink.

5. Paper Stocks Suitable for Covers

The cover of a paperback book takes a lot of punishment. It has to be strong and flexible enough to stand up to being scuffed, bent, folded, and so on. Not only the paper itself, but the image printed on that paper, has to be strong and durable enough to withstand hard use. Many bookstores slap their own stickers on the books. Your covers should be durable enough not to be damaged by the adhesive on those stickers, or by those stickers being peeled off.

It is the image printed on the cover that is the most vulnerable. Because toner can get rubbed off or wear away, unprotected laser toner is not strong enough to serve as the printing on most covers. Ink (such as from an ink-jet) soaks into the paper, and actually get inside the paper fibers. It is thus much more durable and resistant to scuffing. Unfortunately, many ink jet inks are water-soluble, to the point where a reader with sweaty hands could smear the cover.

Whether you use ink or toner, consider using some sort of

varnish or laminate on your cover to protect it, as discussed previously in the section on varnishes and lamination machines. Commercially printed covers almost always have some sort of varnish or transparent overcoat.

Many binding systems use heat to melt glue, and that heat could damage a laminate or varnish.

There is at least one paper out there that comes very close to producing ink-jet covers that need no laminate or varnish. *Mitsubishi Imaging* makes the Diamond Jet line of ink-jet papers that includes one paper called Artist, specially formulated to contend with the problem of water-soluble inks. This photo-grade paper exhibits extremely impressive resistance to water, and prints out very sharp, clear, results. My experiments with this material were most promising. It produces a very sharp looking cover, it binds well, and the stock folds well to form a nice sharp spine edge when folded with the paper grain. Spine edge folding is not as good, but is acceptable, when folded against the grain. I would rate durability as acceptable, but not excellent.

However, the material *feels* strange to some people. It is slick, and yet not slippery. It resists a bit when you slide your finger over it, and even squeaks a bit. At least one person reading a book with a cover made of this material felt the need to wrap the book in something, because he found touching the cover to be unpleasant. However, a cover printed on this stock, and then laminated, would more than durable enough and not have that odd texture. Experiment with it and decide for yourself. Sample pricing of this material: about $75 (lighter card stock semigloss) or $105 (high gloss heavyweight) per box of 100 12" by 18" sheets. Other sizes available, and not all sizes are available in all paper weights. This paper can be hard to find.

I have found that Weyerhaeuser's First Choice Multi-Use 32 pound 94 Super Bright (Weyerhaeuser item #8135 in 11 by 17 sheets) provides very good results with a single-side laminate, but Weyerhaeuser's similar 65-pound stock was too heavy and awkward once laminated.

There are any number of "photo-grade" papers for ink-jet

printers out there, and they do a good job at printing photos. However, these papers are usually priced at something like a dollar a sheet, and are rarely available in anything besides letter-size, which is too small for book covers.

I like Hammermill's JetPrint, a high-quality paper designed to do color ink-jet at less than photo-realistic quality. It costs something less than ten cents a sheet. There are many similar papers, but finding them in the legal sized needed for book-on-demand covers can be difficult. You can cut down the 11 by 17 size of JetPrint down to 8½ by 14 if you can't find the smaller size. Most cover stock is about 80-pound or heavier. JetPrint and the similar papers are only available in 20, 24, and sometimes 28 pound stock weights. These are far too thin and flimsy for use as a cover—until the stock is run through a laminator. When given a single-side laminate these papers make acceptable cover stock.

The similarly named *Jet Print Photo* is a division of International Paper that sells—you'll never guess—a product called Jet Print Photo Paper. It's possible that International bought the brand from Hammermill. Two hundred sheets of 38-pound 11 x 17 go for about $150. Or consider buying the 17 x 22 size and cutting it down to 11 by 17 short-grain. This will put the paper grain parallel to the spine on the final cover. (See the discussion of paper grain on page 69 to see why this is a good thing.)

As we saw in Chapter Four, laser printers, both color and black-and-white, use various forms of toner, which is fused to the page with heat. Toner bonds to the surface of the page, and sits on top of the paper. Ink soaks into paper. Because toner does not penetrate the paper at all, the way ink does, it is far more susceptible to being worn or rubbed off the page than ink. The lettering on the spine of a book with a toner-printed cover will likely flake off over time, simply as a result of the spine being flexed as the book is opened and closed repeatedly.

A company called *Fibermark DSI* makes a line of cover material that is so sophisticated they don't even call it paper. It is a "latex-saturated substrate." The material is called Fibermark ImagEase, and is available in various finishes. It is flexible and durable, and absorbs the toner so completely that it can serve as a paperback cover material without any need for lamination. I have seen and handled a number of paperbacks using this material for the cover, and it works very well. I have tried some samples of this material in my monochrone (i.e. black toner) laser printer, and it did the job just fine. It folds and binds well.

One note of caution: The ImagEase 212 material is rather thick and heavy (as a cover should be). Toner is bonded by heat, and the thicker and denser the paper, the more difficult it is to heat it enough to bond the toner completely. Fibermark DSI advises that ImagEase 212 works properly in certain color printers, but not others, because they don't heat the paper enough.

Do some careful homework, and talk to the folks at Fibermark DSI before buying a color laser printer for use with this material. Most color laser printers should do fine with it, but you wouldn't want to get the one printer on the market that *doesn't* bond its toner well enough to ImagEase 212.

The material is really intended for use in the sort of printers you'd find in a big print shop, rather than for home use, and this bonding issue really applies to such printers. Most office color lasers will likely run slowly enough and hot enough to bond the toner perfectly well.

For the record, Fibermark DSI says that ImagEase works well in a Xerox DocuColor 70, but that a Xerox DocuColor 40 would not heat the paper enough to insure a solid bond. The image will bond in a DocuColor 40, but you'll be able to scrape it off with your fingernail. It won't be strong enough for a cover. A gentleman from Xerox suggested that the problem with the DocuColor 40 could be solved simply by briefly reheating the 212-weight paper (sorry, latex-saturated substrate) in a household oven. Fibermark advises that the Xerox Regal 5790 (which runs more slowly than the similar DocuColor 40, giving the heat more time to penetrate) or the Canon CLC 700 or 800, or any of the Ricoh color printers, would provide a strong toner bond. (Upon request, Fibermark DSI will send a detailed chart on printer compatibility which goes into this in more detail.)

An alternate solution might be to go to the Fibermark ImagEase 135. (The numbers 212 and 135 refer to the weight of the paper in grams per square meter, or gsm, the metric measure of paper weight.) As it is thinner and lighter, it will require less heating to get a good solid bond to the toner.

Rough pricing of this material: $220 for 1,000 sheets of 12 by 18 inch paper. Fibermark prefers to sell a minimum of 5,000 sheets. Smaller sizes are available, with pricing more or less in line with the above, on a price per square-inch basis. They will cut custom sizes, but want to sell 10,000 sheets at a clip. They are willing to ship generous evaluation samples.

Even if you don't have a color laser printer, ImagEase could be the way to go. It would be great for out-sourced covers. You could bring a stack of ImagEase to Kinko's, or whoever, and have them print your covers on that stock.

(Notes: The company that made this material, Rexam DSI, was purchased by Fibermark. As of the current date, the Fibermark website contains no information on ImagEase, but a phone call confirmed the material is still available.)

In similar vein a company called *ICG/Holliston* makes a linen substrate that can be used as the cover material for the case binding on hardcover books, without any need for lamination. It works very well with toner-based printing.

As the above discussion should make plain, there is no one perfect way to do covers. There are instead lots of fairly good choices, all of which have drawbacks as well as advantages. Do your homework, and do some test runs, before you commit yourself to anything.

Now that we've looked at what a cover's made of, let's talk about what goes on it.

6. Cover Size and Layout

As we have seen, the paper for a cover for a wraparound book cover must be long enough to go completely around the front, spine, and back of a book. For a half-letter sized book (5½ inch wide pages) book with a one-inch spine, that means stock at least 12 inches long. The easiest thing to do it to print covers on legal-size stock, which is 14 inches long. Be sure that any printer you get for doing covers can handle legal-size paper. Some of the newer ink jets can do "banners," meaning more or less infinitely long pieces of paper 8½ inches wide. Obviously, they could do legal and then some. Such a printer might allow you more flexibility in your covers.

You'll want quite literally to allow yourself some margin for error on your covers, as you'll be trimming away a small amount of all the edges of your covers. Don't have vital design elements right at the edge of the paper. Many printers won't print to the edge of the sheet anyway. If your printer only prints to, say, within an eighth of an inch of the paper's edge, you'll need to allow at least another eighth or quarter of an inch inside that limit to give yourself some wiggle room.

If you print books with full letter-sized paper, and want to do wraparound covers, you'll need a printer that can handle larger paper. To do a wraparound cover of an inch-wide letter-size book, you'll need cover stock at least 11 by 18—and that's not a standard size by any means. You could find larger stock and cut it down, but that would be a nuisance and involve a lot of wasted paper. Probably the easiest thing to do would be to design your book with margins suitable for, say, a 8¼ inch trim width (that is, final size after the last trim) and use 11 by 17 stock for the cover. Ink-jet printers that can do 11 by 17 paper are falling in price, and some are available for $400. That's cheap enough that you might as well get the larger printer. Even if you don't make covers that big, you'll be able to do posters and cover blow-ups and other promotional items.

Whatever printer you get for covers, make sure it handles your cover-stock gracefully. Bring samples of the cover stock you want to use along to the computer store and test the paper-handling ability of the printer that interests you.

Once you get your printer home, but before you get too deeply involved with a complex cover layout, do a test layout and make sure your printer can handle the job. Lay out a rough cover design that is as simple as possible, while including the basic elements. Once you have a cover rough, use your word processor or page layout software to draw a line exactly down the center of your cover's spine as seen on the screen. Adjust the spine lettering so that it exactly bisects this line running down the edge of the spine.

Create a second page to serve as the inside of this test cover. Once again, place a line smack down the center of where the spine will be. Figure out where your scoring lines will be, and mark them on either side of the center line. (You might or might not want to print the scoring lines on your final production covers, but for testing purposes, at least, it is helpful to see where they will go.)

Now print your outside cover, using the cover stock you plan to use for the final cover. When the cover comes out of the printer, turn it over so the same end goes in first as on the first pass, and put it back in the printer. Print the inside cover. (Or print the inside cover first. Just be sure you do it the same way every time, and be careful about what end goes through the printer first. Consistency helps you avoid muddle.)

Use a ruler and see how close the center lines on both sides are to actually being in the center, and to each other. Hold the cover up to the light and see if the spine lettering on the outside is centered between the scoring marks. Run off two or three such covers and compare between them for consistency.

What you want to know if the printer is spot-on every time, if it produces consistently inaccurate results, or if the covers come out misaligned a different way each time. You also want to know how great your margin of error is.

If the printer aligns everything perfectly, congratulations. If it consistently prints the center line a half inch off from where it should be, simply adjust all the cover elements by a half inch. If, however, each and every cover comes out cockeyed in a different way, your printer is probably not up to handling your cover stock. You'll either have to shift to another paper (probably a lighter one) or have someone else print your covers.

If you discover that the covers print out more or less all right every time, but that there is a certain amount of slopping about from one copy to the next, maybe with the copy moving a quarter-inch to the right one time and an eighth of an inch to the left the next, you can design a cover that will accommodate that problem. Use smaller spine lettering so it won't get caught by the cover fold. Don't plan on having two contrasting colors meet exactly on the edge of the spine. Avoid putting vital cover elements close to any outer edges, in case they get trimmed off due to changes in the alignment of the cover from one copy to the next.

My covers for this book are about 12 inches by 8½ inches. I print that cover image on a 17 by 11 sheet, and print it with a box around the cover image itself. If the paper comes out of the printer a bit crooked, it doesn't matter. I cut around the edge of the box, and I have a squared-off sheet with the cover precisely aligned on it. When I run the finished book through the trimmer, the box is trimmed away.

However you do your printing, it is far better to get all the details worked out early on, rather than getting nailed by surprises later on. Establish orderly procedures that take into account how well your hardware works, and allows for its limitations. And, of course, once you have everything worked out for one title, it should be easy enough to set up a generic, template cover upon which to base covers for future books.

One last thought: if you decide to go ahead and print on the inside cover anyway, so as to put down the scoring lines, think about printing some about-the-book or about-the-author text, and maybe even an author photo (if your printer can do it properly) on the inside of the cover as well.

No matter how you do your covers, be prepared to experiment a lot before you get a cover that works for you. You'll use up some paper before you get it right.

Now we know how to lay out the pages and print them, and how to make up the book covers. Let's get started on the many ways of putting those pieces of paper into books. Let's talk binding.

Chapter Six
Hand Binding and "Office" Thermal Binders

Now we come to the heart of the matter, and the part that, at least to me, seemed the most difficult and mysterious part of the job of making books. How did they get the paper to stick together edge-on like that? How did they turn stacks of paper into books?

In this chapter and the one that follows, we will discuss binding procedures that produce real paperback books: pages bound together inside a custom wraparound paper cover with a flexible, printed, spine.

There are many, many ways of binding books that don't match this description, from simply stapling a few pages together, to spiral-binding and velobinding, and so on. But holding pages together is not enough. There aren't many velobound books on sale at the local bookshop, and precious few spiral-bound volumes at the local library. This guide is for people who want to make books that look like books.

In this chapter, we will discuss binding books using nothing but hand tools and objects commonly found about the home, and binding using small-scale "office" thermal binders. In the next chapter, we'll cover production binding machines.

This book, and these chapters, won't talk much about hardcover books. The goal here is the small-scale mass production of books, one book, five books, ten books, a hundred books at a time. Hardcover book-making requires so much hand-labor that such small-scale mass-production becomes impractical.

I have left out information on making hardcovers for another reason. My goal was to concentrate on information that was hard for me to find. There are plenty of how-to books on making hardcover books available. *Dover Publications* offers quite a number of them.

Be advised that most of the titles on this subject are reprints of decades-old texts, so check the copyright dates. The processes haven't changed much, but more than likely most of the companies these books list as sources of supply are long gone. *LBS (Library Binding Service) Bookmaker's, Hollander's* and *Talas* are current sources of supplies and information on hardcover binding. See the *Names and Numbers* section at the back of the book. Don Lancaster presents a bibliography of titles on the subject of hardcover book making on his web site, *www.tinaja.com.*

Let me make one note in passing on hardcovers, however. In the old days, hardcover, or case-bound, books were always sewn together, typically by a technique known as Smythe-sewing. The pages were held together not by glue, but by thread. Some books are still made this way, but these days, the majority of hardcover books have the pages glued together in precisely the same way as we'll discuss in the section on production binding machines. This assembled set of pages, sometimes called a "book block" is then glued into a case binding. Sometimes the pages are glued right to the spine. Sometimes they are glued to the front and back covers, in much the same way as the layflat binding described below. It used to be that a hardcover was substantially different than a paperback. Not anymore. In essence, the average modern hardcover is a paperback book glued into a set of stiff covers.

You could use any of the techniques discussed in this book to bind the pages together, then make up your own case binding and glue your pages into that binding, making a hardcover book. But there's a lot of labor involved in such a project, and it's beyond the scope of this book. See Evans' *Book-on-Demand Publishing* for a full discussion of this technique. See the section in Chapter Seven on *Maping*'s *FastBind* system for a brief mention of their system for hand-making hard-cover books.

All that to one side, let's concentrate on the bread-and-butter of home book-on-demand: trade-size paperback books.

A quick explanatory note: paperback books that are roughly the size of normal hardcover books are called "trade" or "trade-size" paperback. The smaller, pocketbook sized paperbacks are called "mass-market." There's no reason the techniques described below could not be used for smaller books, but for various reasons, trade-size will probably be the best way to go for most home publishers. It will cost the home book-binder just about the same per unit to make a mass-market size, but, because it will look like a smaller (and cheaper) book, it will be hard to get a price for it that will cover the costs of making it. Looking at it another way, the larger the book-page, the more words per page, and the fewer pages (and less paper) needed. Thus it will cost you less to make a larger book, and you'll be able to charge more for it.

Doing Product Research

It took me a long time to figure out that binding books together is not anywhere near as difficult or mysterious as I thought—or, more accurately, as I allowed myself to be convinced it was. I thought glues were not as good as they are, and that bonding pieces of paper together edge-on was a tremendous technical challenge for the basement binder. I was wrong on both points.

I didn't gather enough information before I went ahead. I didn't investigate all the options. I made the mistake of making a fairly substantial investment in a binding system based on pretty scanty information. Don't let that happen to you. Do research. Do homework. Use this guide, not as a endpoint, but as a starting point for your investigations. Use the information I'm providing, but don't trust it too far. I can make mistakes, you might not agree with my opinions, you might find information I could not, and, of course, things change. Find things out for yourself.

Before you invest in *any* commercial hardware, get your hands on samples of books bound with the process in question. The manufacturer or local distributor should be able to send you copies, or ought to be able to give you a list of local print-shops that use the machine you're interested in. Alternately, you might try working the phone to locate that information. Check the Internet. Lots of print-shops have web sites, and lots of them list the equipment they use.

Print shops like Kinko's often have a binding machine in the back, though these machines are used mostly for padding (that a lot of blind alleys, the worst of which was getting stuck with a short-run "office" binding system that just didn't make the grade.

Let's discuss these "office" systems first, and get them out of the way.

2. Short-Run Office Thermal Binding Systems

The blind alley I went down was in having faith in a commercial product that got good reviews but only deserved fair ones: a binding system called Penta-Bind, made by *Unibind.* Penta-Bind has since been discontinued, which illustrates another mistake I made: failing to confirm that my source of supply for binding materials would be reliable.

Penta-Bind had one unique feature (which I'll discuss in a minute) that got me interested in it, but it was in many ways a pretty standard thermal office binding system, closely comparable to all the others on the market. As the basics of the Penta-Bind system pertain to all the office thermal binding systems, I'll describe it in some detail.

Penta-Bind was made by a European company called *Unibind.* They seem to have offices in Belgium, Holland, and maybe Italy, though I'm not sure where the home office is. I believe it is in Belgium. (Many of the short-run office binding systems are European. Even many of the ones that present themselves as American are really just relabeling and distributing European products. Nothing wrong with that, but a little truth in advertising wouldn't hurt.)

In order to bind a book with the Penta-Bind system, you dropped the pages into what Unibind called a binding insert. The insert consisted of two pieces of parchment-like paper glued edge-on to either side of a strip of adhesive that melted under heat. This strip of thermal glue was what bound the book together. The binding inserts came in various widths from about an eight of an inch to an inch. The glue on the spine edge of the binding insert, at room temperature, had the consistency of a rubbery, flexible, plastic (which is, basically, what it was). The binding sleeve, with the pages in it, went inside a paper book cover, which could be any sort of paper you liked, printed however you liked, and scored to the proper width along the spine.

You then dropped the whole affair into the official Penta-Bind Binding Machine, which heated the glue, melting it, causing it to adhere both to the pages and to the cover.

The trouble was that, for me at least, it generally worked *nearly* well enough to hold the pages to the cover, but only nearly. Maybe the heater on my particular unit was adjusted too low. Maybe there is some meaning in the fact that the Penta-Bind system is no longer with us. But lots of my bindings fell apart, and lots of them had pages that fell out.

The other annoying thing was that the official Penta-Bind Binding Machine was nothing more or less than a heater with a non-stick pad the right size and shape to hold a book spine-down, plus a thermostat with an LED indicator and a little gizmo that went BEEP! when the heater reached temperature, and another gizmo that used an electric eye to spot a book being dropped into the binder and started a timer that went BEEP! when the binding cycle was done. As it usually took at least two binding cycles to get the pages to stick, this feature was likewise of limited value.

I paid several hundred dollars for the heater, and several hundred more for binding inserts that more or less worked and were not always the right width, and were always the wrong length. They were sized for binding letter-sized paper (actually, A4-sized, that being a metric size a bit longer and thinner than letter-size paper) on the long edge. Because I was doing half-letter sized books, I had to slice away a couple of inches of each binding insert.

This description of Penta-Bind, with the exception of one detail, generally sums up all the low-end (under $1,000, most under $600) short-run office thermal-binding systems: A heating unit with no moving parts to speak of melts a thermal glue placed between the edge of the book pages and the spine of the outer cover, causing the glue to stick to the pages and the cover.

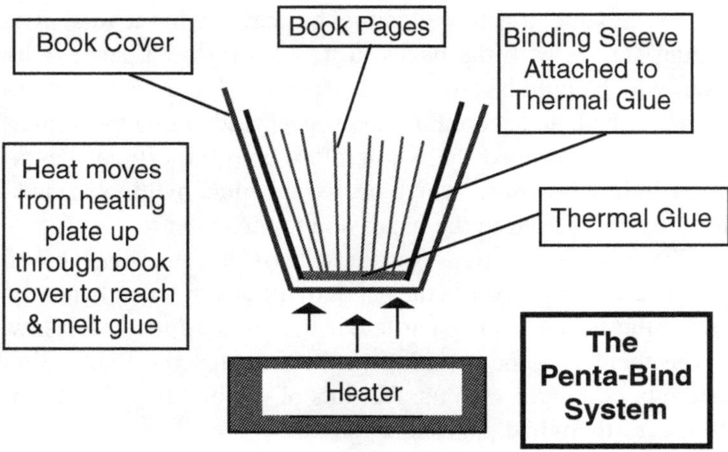

The one exception had to do with the one way in which Penta-Bind was better than the other low-end systems. It allowed you to print your own covers. Most of the other low-end binding systems, which are still available, want you to use prefabricated covers provided by the people who sold you the binder.

These covers have the thermal adhesive already in the spine. They come in a limited number of sizes. Nearly all of the brands make covers for letter-sized paper only. Generally speaking, you have to order your custom-printed covers in lots of something like fifty or a hundred. If you have five different titles and want to do twenty of each, you have to buy a hundred of each cover and be stuck with eighty leftovers. Many of the binding system companies won't even do spine printing for you, let alone allow you to do it for yourself.

None of these systems is more than a heater with a thermostat and maybe a timer. A few up toward and beyond the $1,000 range have features such as a heating plate that rocks back and forth a little, jogging the paper to make sure it comes in contact with the adhesive. For $1,000 I can wiggle the pages back and forth myself.

Virtually all of these systems are as interested in getting you to buy their special, advanced, high-tech, custom-made covers (on which they make a lot of money) as they are interested in getting you to buy their binder. It's sort of a system of mini-monopolies: each bindery-system seller wants to convince you that only their covers will work in their machine. (I might add that, in my somewhat limited experience, these companies act like monopolies, with all the alert customer service, rapid response, and high degree of competence one might expect from, say, the Soviet Ministry of Agriculture. Trying to track down what happened to Penta-Bind was a lot of fun. I never did get entirely clear information, which is in and of itself pretty suggestive.)

All of these companies say that you MUST use only their brand of binding cover or your warranty will be voided, the skies will fall, and all your bindings will explode. Just because they want you to believe this sort of talk doesn't mean it's true. Unless brand X uses a glue that melts at 350 degrees but disintegrates at 375, while brand Y melts at 300 and evaporates at 325, which makes no sense, how could it be true?

As with the official Pillsbury cookbooks that call for you to use only Pillsbury© Brand® Bread$_{tm}$ Flour© in all the recipes, you start to get the impression they are pressing a bit too hard to make their point. Brand X covers should work just fine in brand Y heaters, assuming they heat to more or less the same temperature. A heater is a heater. Nor are the glues in these covers anything special. As we'll see, the adhesives used in true perfect binder have to be liquid at some temperatures, retain the proper viscosity, and be transported from a glue pot to a moving book block. The office thermal binder glues just need to melt and then cool. It doesn't do anything mechanically, and so it can't harm much of anything.

The makers of the office thermal binders want you locked in to their systems, but their machines don't suit the needs of a home book-on-demand operation. These systems are pretty obviously aimed at the small-business office, where they might actually make sense, and where you can sell a $40 heater for $400 if you call it high-tech office equipment.

These aren't bad products. They work, and they do what they set out to do. But what they do is not what you want to do. The products are simply a poor fit for the book-on-demand printer.

You want to print any size book you want, with any width and length of spine you want, and you want to be able to put different printing on all your covers.

They want all your books to have eleven-inch long spines and pages 8½ inches wide, to have spine widths that closely conform to what they feel like manufacturing, and to have you utterly dependent on them for the printing on the cover, most likely selling you a hundred covers at a clip.

You don't have to play along. None of these systems is anything more than a way to melt glue to bond pages to covers. As we'll see in a minute, there's more than one way to do that.

The fact that Unibind has dropped its Penta-Bind, sort-of-good for book-on-demand, system, and yet is still in business selling variations on the systems described immediately above, suggests there's a market for those binders, but not for the kind of binding system book-on-demand printers need. However, if and when a large home-office book-on-demand market develops, the prices on these heating units will drop like a stone and there will be a whole flurry of breakthrough discoveries that allow you print your own covers and bind them in the official Bind-a-Tronic 3000, using the new miracle ingredient, glue. In the meantime, you have other choices.

3. The Evans Do-It-Yourself Thermal Binding Systems

In his book, *Book-On-Demand Publishing*, Rupert Evans demonstrates that a ordinary $40 electric skillet will do the job of heating glue just as well as the commercial systems described above. It sounds absurd, but I can say from experience that there is precious little of value that my binding machine could do that an electric skillet couldn't provide. In fact, an electric skillet has an adjustable thermostat—something my "real" binding machine lacked. With a thermostat-controlled heater, it is possible to crank up the temperature if the glue isn't melting completely. Without an adjustable thermostat on my binder, I was out of luck.

In order to use an electric skillet for a binding machine, all that is required is to rig up a way to hold up the book in the proper position—spine down on the griddle—while the binding is being heated. An exotic material called "wood" can be used to build this item. This support should be adjustable to allow for different widths of book.

Basically you need two pieces of wood about four or five inches high and ten or twelve inches long, held together with long carriage bolts run through drilled holes, with butterfly nuts to do the adjusting. The support should not be tight against the book, but loose enough that you can move the pages around a bit by hand as the glue heats. This ensures that the pages fan out enough so that all of them get glued. Evans suggests that you avoid pine (the heat could cause turpentine to drip out of it), and use a fir-based plywood to build the support.

Lay down release paper, waxy side up, on the griddle top. Evans suggests a binding temperature of 350 degrees Fahrenheit, giving the book about thirty seconds to bind. Put the book in the support, let it sit there for thirty seconds, perhaps wiggling the pages a bit to get them in contact with the glue, and then set to one side, spine down, and give it time to cool.

Note that those are basically the instructions for my Penta-Bind System: a further demonstration that a heater is a heater.

So much for the commercial binding machine. What about what about the commercial binding adhesive? As noted, the Penta-Bind system is no more, so you can't use that, and just as well. However, here too Rupert Evans has come up with ways to eliminate the need for the commercial products; in fact, two ways.

The diagram shows a rough idea of the first technique. Jog the pages of your book so the spine edges of the paper are properly lined up. Put the pages of a book (or several books, with different-color waste paper, or strips of release paper, between them) between two boards, with the spine edge exposed. Weight or clamp the boards to hold the stack together. Use a high-temperature glue-gun with at least forty watts' power, and a glue-

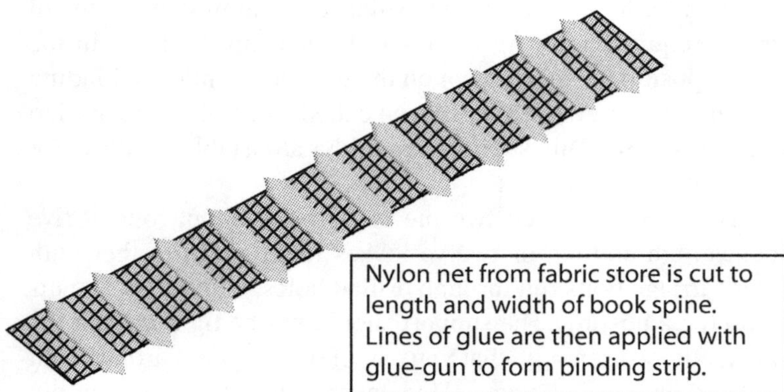

outlet hole no larger than about $^1/_{16}$th inch in diameter. Run lines of glue along the spine edge as shown, with the lines about a half-inch apart, and with a half inch of space between the end of the page and the first line of glue. (The glue melts in the binding machine, and fills in the empty spaces. Room is left at the top and bottom of the spine to keep too much glue from melting out the ends of the spine.)

Nylon net from fabric store is cut to length and width of book spine. Lines of glue are then applied with glue-gun to form binding strip.

Trim away any excess glue with a sharp knife. If you have run glue over several sets of book pages at once, cut the page sets apart, using the scrap material between them to guide you. Drop the glued set of pages into a scored cover, and use a commercial or home-made binding machine, as described above, to remelt the glue, binding the book.

I have only done a few experiments with this process, but on my first try, this technique yielded a stronger and more flexible binding than I ever got with Penta-Bind.

Here's Evans' second technique, closely related to the first: get a sheet of fine nylon net, sometimes known as "horsehair." It's a sort of reinforcing material used in sewing, and it is available from fabric stores. Lay it out on a non-stick surface and run lines of glue (again, a half-inch apart) from a glue gun along it. Allow to cool. Store until needed. Cut strips of this material to

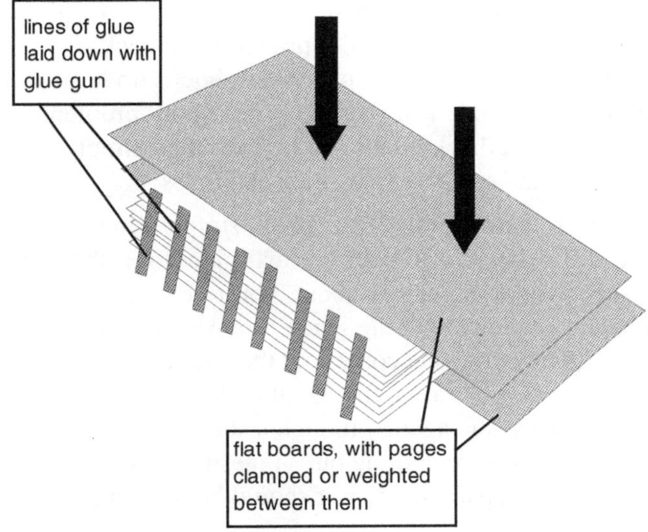

the width of the spine of your book, and about a half-inch shorter that the spine. Place the strip of horsehair in the scored spine of the book cover, place the book pages on top of the strip, and put the whole thing in the thermal binder. I have not tested this technique, but I see no reason it wouldn't work.

As if I had not made it clear enough already, virtually all of the do-it-yourself information above is from Rupert Evans' book, *Book-On-Demand Publishing.*

I have presented a much foreshortened version of Evans' ideas. While I believe I have presented a useful summary here, there is a great deal more information on all these techniques in his book, and he discusses a whole range of more elaborate techniques as well. If you are interested in hand-binding books, you should buy and read his book.

4. Using Your Binder For Instant Paperback Repair

Before moving on to our next main topic, I'm going to indulge in a slight digression at this point, and pass along a few tips not directly related to book-on-demand printing. A commercial or homemade thermal binding machine of the sort discussed above can be use to repair commercially printed paperbacks.

Nearly every commercial paperback book, trade or mass-market size, is bound with thermal glue. Many problems suffered by such paperback books can be solved by simply reheating the glue in the spine. A few quick examples:

- If pages are coming out of the book, simply drop the whole book, spine down, into your thermal binder for about thirty seconds. Wiggle the pages about a bit to encourage them to get back in contact with the glue. Remove and allow to cool. Scrape off any glue that comes out the ends of the binding. In a large number of cases, this is all that is needed to get the pages stuck back in again—though it should be noted that the fact that pages were coming out suggests the book was fragile before the repair, and that it should be treated carefully.

- If the spine is badly rounded-in, as often happens to thick books that get hard reading (especially in the bathtub), the situation can often be improved, if not resolved all together. Heat the spine, then, while the glue is still hot and pliable, press in on the front and back covers and down on the spine-opposite edge to force the spine back out toward squareness. Put the book between weighted or clamped boards for a while after it cools to encourage it to stay square.

- If the spine or cover itself is torn or broken, you can probably peel it away from the book after a run or two through the binder. Scrape away the glue that wants to come off easily. The glue that wants to stay will help form the new bind. Print and score a new cover and rebind the pages into the new cover with one of the above techniques.

I should note that none of these are remotely archival techniques, and aren't the sort of thing one should do to a family heirloom. I would expect that it would be fairly easy to weaken a binding by reheating too much, or too often. Work carefully. You run a risk of making things worse. However, I have used all these tricks to extend the life of much-read books that have seen better days. These tricks might turn an archivist's hair green, but they work.

5. Cold Glue Binding

Cold glue might better be termed room-temperature glue. The term distinguishes these books from the hot-glue techniques described above, and isn't meant to suggest these glues dry in the refrigerator. "Air-drying" would be a more accurate description, but "cold glue" sounds better, and is the term in general use. Some of the glues used, but not all, are water-based, and can therefore be dissolved in water.

Cold glues have advantages and disadvantages when compared to thermal glues. The big drawback is drying time. A thermal binding takes only a few minutes, or even seconds, to cool, and then it's ready to go. A cold-glue binding might take up to several hours to dry. Cold glues often require several coats of adhesive, which means you have to go back every x period of time to slap on the second or third coat. It's a more labor-intensive process.

On the up side, cold glues can (but don't necessarily) provide a stronger bond. Another advantage is that cold glues tend to migrate into the paper, and that means these glues can be used on papers—such as coated stock—that don't adhere well using thermal glues.

But perhaps the most important advantage is that cold-glue binding doesn't take a lot of hardware. Leaving paper-cutting out of it for the moment, the minimum equipment list for small-scale cold-glue binding is:

(a) two boards,

(b) weights or clamps, and

(c) a cheap paintbrush.

Aside from the cost of a pot of glue, five bucks (ten if you get fancier clamps) turns you into a bookbinder.

A very quick summing up of cold-glue binding for a paperback book would be as follows:

- Jog the pages so that the spine-edge is squared up. Put the pages between two boards, with the spine edge peeking out, and weight or clamp in position.

- Paint the spine edge with two or three thin coats of glue. Allow the glue to dry between coats. For extra strength, take a piece of muslin or other cloth the length of the spine and a half inch or so wider than the spine. Lay it over the last coat of glue. Wrap the excess cloth around the front and back covers and glue it in place, or trim it away square with the edge of the spine. The pages of the book are now assembled and attached to each other to form a book block.

- Allow to dry. Once dry, apply water-based or thermal glue to the inside spine edge of your scored book cover, and glue it to the spine of your book block. Allow to dry or cool.

- Alternately, use the slightly more complicated lay-flat binding procedure discussed below.

There are a number of possible variations on this procedure, but that is basically that.

One variant is to do a lay-flat binding. If you used an piece of muslin or other reinforcing cloth that hung off the front and back edges of the spine by an inch or so, you could then glue the muslin to the front and back covers, while leaving the spine of the book block unattached to the cover spine. The front and back pages of the book will take up some of the glue holding the muslin to the cover, and would therefore bond to them, and not open fully, but the rest of the book pages should open very easily. If you add a blank page to the front and back of your book, the cover page and last page of the text won't be glued to the cover in this rather awkward manner.

The diagram shows the two possible positions for the glue holding the spine to the book-blocks. Note that the second example is scored to bend, not just at the spine, but just past the

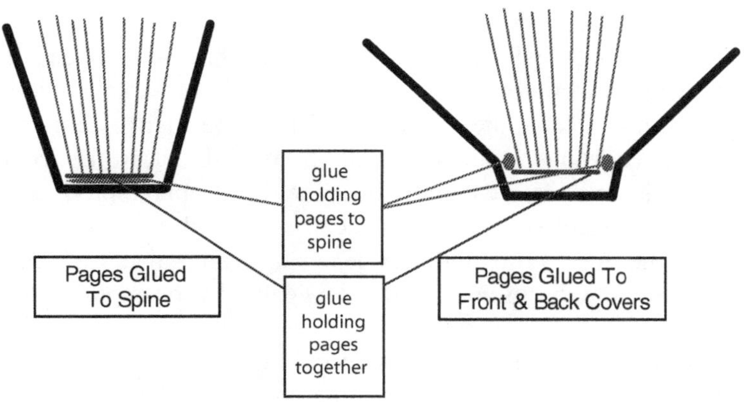

two lines of glue attaching it to the front and back covers. This second pair of scores lets the spine open more freely.

I have found that I can get away with not using the cloth reinforcing strip to form the lay-flay binding. I have had good results by instead gluing the cloth to the spine and running a line of thermal glue between the back-cover spine-fold score and the lay-flat score, and pressing the properly positioned and oriented book-block onto the line of glue. Once that is set (a matter of seconds) I run another line of thermal glue between the front cover scores, and fold it carefully over the front of the book.

The obvious question at this point is, *what sort of glue should I use?* Yellow carpenter's glue is not a good choice, and Elmer's white glue is a worse one, as these glues are not really flexible enough for service in a binding. All accounts I have read so far sing the praises of a class of adhesives known collectively as PVAs (poly-vinyl acetate). (Elmer's is in fact a PVA, but it is not the right sort of PVA.) Evans says that an adhesive from Wisdom Adhesives, #150F, is the best for binding. Evans advises one coat of the #150F diluted with a small amount of water, and a second undiluted coat applied about five minutes later. Allow the second coat to dry for at least an hour.

There are various other PVA-class binding glues available. See the bookbinding and library supply vendors listed in the

Names and Numbers section.

I have not tried Wisdom's #150F, but I have used Aleene's Tacky Glue (available at craft stores). It works remarkably well, and affords a far more flexible bind than the thermal-binding techniques I have used. By gluing the book-block to the front and back cover, rather than to the spine, I was able to produce a book with excellent lay-flat characteristics, with a strong, flexible spine. And, I might add, Aleene's dries in a matter of minutes. (See the discussion below on Gigabooks for another glue choice.)

If you're going to do a lot of cold-glue binding, consider buying or making yourself a padding press. A padding press allows you to stack paper spine-edge in, flat against a guide-board, so that all the spine edges are flush up against the guide and exactly even with each other. The stack of paper is then clamped together, and then swung around to expose the spine edges, so that they can be painted with glue. Once the last coat of glue is dried, you remove the pages from the press, and carefully cut the page sets apart. Even a small padding press will allow you to stack your pages a foot high, allowing you to bind page-sets for several books at once. *Printers' Shopper*, *MachineRunner*, and *Factory-Express* all sell various models of padding presses.

Fan-gluing is a technique that improves the quality of cold-glue binding. Basically, the sheets are clamped together with the spine side clear of the clamps, so that if they were held vertically, they might flop over a bit to one side, fanning out the edges of the paper. By applying the glue while more of each sheet is exposed, and then squaring the sheets back up, glue gets in between the sheets, making for a strong and flexible bind. *TeMPeR Productions* makes a fan-gluing press for about $450 that is designed to make this binding process more precise and repeatable. It's more or less similar to the padding press described above, except the paper sheets are rotated into a vertical position for glue application, and the pages are clamped some way in from the edge to give the sheets enough free play to fan properly.

Planax (represented by Import-Graphics in the U.S.) sells

the DF 83 Fanning Binder for about $2,700. It works in a similar fashion to the TeMPeR unit, but it is made of steel, not wood, has more adjustability, with some additional features, and can handle thicker paper. But even so, that's a lot to spend on a sophisticated clamping system.

The procedures for cold-glue binding are simple enough, and the possible variants wide enough, that you should spend some time experimenting, until you find out what works for you.

6. Side-Stapling

This technique is almost too easy. Grab yourself a copy of National Geographic and run your fingers along the front and back edges of the cover just by the spine. You'll feel small ridges in the cover, parallel to the spine. Those are staples. You can buy a heavy-duty stapler that will punch through up to 250 sheets for somewhere between $70 and $100. Printers' Shopper sells several such staplers. Collect your pages, staple them two or three times along the spine edge, and glue a wrap-around cover around the bound pages as discussed above. If it's good enough for the Geographic, it's good enough for you. There are obvious limits to this technique. Thicker books won't work, for example. Still, it certainly has the advantage of simplicity.

7. Gigabooks

Finally, I'll close this section by mentioning one other source of information on hand-binding paperbacks: *Gigabooks*. This is a small company that sells how-to books, a line of roughly $100-$200 kits for hand-binding paperback books, and various supplies for book-binding. Their hardware line consists of better-than-home-made clamps and gadgets for doing hand binding basically as described above. Their two titles are entitled *The Perfect Binding Handbook*, and *Easy Hardcover Bookbinding: The Booklet.* The booklet on hardcover tells how to do hardcover binding by hand.

The Perfect Binding Handbook is useful and informative, but quite limited in scope. That being said, I would recommend the book to anyone looking into techniques for doing hand binding, and I'd tell such persons to consider buying the Gigabook presses as well.

The main flaw with the book is that it's exclusively focused on one somewhat specialized binding technique that is really only suitable for very limited numbers of books. The book does give an excellent overview of the hand-binding presses that Gigabooks sells, and gives you the chance to make a well-informed buying decision about the presses. The book even gives instructions on how to make the presses yourself. However, unless you have a better home workshop than I do, it's probably not worth building your own when Chet Novicki, the owner of Gigabooks, is willing to sell you a ready-to-go press at a reasonable price.

The book itself is made by the author, and serves as a good demonstration of how well his system works. The binding itself is very good. The page design is on the amateur side, but that's a quibble, a design choices I disagree with, rather than problem with Mr. Novicki's binding procedure.

The way Mr. Novicki makes books bears a strong family resemblance to the cold-glue hand-binding techniques we have discussed already. However, there are a few items of interest about Mr. Novicki's technique that I would like to mention. First is that he suggests the use of plain old contact cement as the binding adhesive. He discusses the merit of this brand over that, but concludes that just about any contact adhesive would do. It obviously works, as evidenced by the binding on the book itself, but it's the first I'd ever heard of using contact cement for book-binding.

Second is his technique for avoiding paper-cutting altogether (except for trimming the covers). He does thos printing the book in a sequence of four-page signatures, which he folds one at time by hand, then stacks together in the proper order. He then glues the folded edges of the stack sheets together. The first signature of pages 1, 2, 3, 4 (see diagram, page 80) is folded, then signature two of pages 5, 6, 7, 8 is folded and placed on top of the first signature, and so on. It seems to me that this introduces a lot of hand work. If you're doing few enough books, and they are short enough that the fold-one-page-at-a-time trick is practical, then you probably have few enough pages to deal with so that you could just cut the pages ten at a time on an office paper trimmer. I have seen two versions of Mr. Novicki's book—one with cut, and one with folded pages. He seems of two minds about the issue as well. On the other hand, the folding technique works, and it lets you avoid cutting, and those are pretty telling argument. Take a look at the Gigabooks site, and see what you think.

Rupert Evans told me about a very old Chinese technique, similar to the Gigabooks procedure, that not only eliminates paper-cutting—it eliminates the need for double-sided printing. The pages are printed two-up, as described above, but on a single sheet of paper, with one side left blank. The pages are folded and stacked in order, printed side out, and then glued together on the opposite side of the sheet from the fold. In other words, if the folded sides of the stack sheets are on the right, the left-hand sides of the folded sheets are glued together.

Obviously, this procedure requires the use of thin paper to make up for the doubled-up sheets, and obviously the books are going to look a little odd, in a way that might confuse some readers. But there must be some merit in a technique that cuts out both duplexing and paper-cutting.

That concludes our quick survey of a number of easy way to bind those pages. To my way of thinking, the biggest secret of hand-binding paperback books is that there is no secret to it. There's lots of ways to do it. You should be able to pick and choose between the various techniques we've described, and set up a production procedure that works for you.

The trouble is that, while hand-binding is straightforward enough that anyone can do it, most people will find it to be slow going. It will suit for very short production runs. Few book printers doing longer runs will have the patience to bind fifty or a hundred books one at a time by hand, and/or will conclude that it's not the best use of their time.

In the next chapter, we'll talk about some heavier-duty hardware that will make the job go faster.

Chapter Seven
Small Binding Machines

The binding techniques we have discussed thus far are suitable for extremely small-scale and short-term operations. If you want to print and bind five books some Saturday morning, the information provided thus far should tell you all you need to know.

However, there is some production level at which all the handwork becomes counterproductive. Even leaving the nuisance and drudgery and repetitive labor out of the equation, if it takes, for example, ten minutes of labor to bind each book, that's only six books an hour, or forty-eight in an eight-hour day.

It is also relatively skilled labor that requires a bit of practice. Given the complexity of the process, it's likely that your books will vary in binding rigidity and strength, adhesive thickness, and so on, from batch to batch, or even book to book. A binding system that required nothing more than pushing a button and throwing a lever, and that provided more consistent results, would have a lot going for it.

1. Automated Processes and Hand-Labor

The processes required to create books on demand can be divided into three categories:

- Labor-intensive initial creation (writing, editing, laying-out, proofing, etc.).

- Highly automated repetitive processes (printing out multiple identical copies).

- Jobs that must be done repeatedly, one at a time, by hand (scoring, cutting, gluing, trimming, etc.).\

Let's consider this in more detail. Writing the book, laying it out, designing the cover, and so on, are all very much labor-intensive tasks. However, they are jobs that only need to be done once. When your text is written and edited and proofread, and your layout is perfected, and your cover is the way you want it, you're done, aside from updates and fixing typos.

The task of printing out copies of your book will be more or less automated. A good fast duplexing laser can kick out a copy of a shorter book in five or ten minutes, and a modern ink-jet won't take more than a minute or two to run off the color cover.

You can even leave the printers printing while you go do something else. With a big enough paper bin, a printer could run unattended for hours. Thus, even if it takes time for the printer to crank out pages, it is not labor-intensive time. Assuming that all the machines are behaving, all you have to do is check in and reload paper, toner, and ink now and then. (Here is another way having a duplex printer become important. Printing the front and back of ten or twenty copies of a book in separate passes on a simplex printer will quickly turn into a nightmare that offers vast opportunities for mistakes and confusion.)

If you have a duplex printer and your printing routine is straightforward enough, you might well want to do what I do: turn the computer and printer on when you're doing something else, and set them to work. Stockpile page-sets for five or ten or twenty or fifty copies of your book, printing ten before dinner and twenty the next evening. You could accumulate all your page sets over time, and then get the binding done all in one go, rather than binding them as soon as the pages are printed. (This also gives you a chance to stack the page-sets under weights, encouraging them to flatten. The flatter the pages are when you bind them, the better. Pages that curl apart from each other look bad, and can weaken the binding enough to break the spine.)

So, until the binding process, all the tasks are either do-it-once jobs (like writing the book or laying it out) or largely automated, if repetitive, jobs (like pushing one button to print the entire book or even multiple copies of the book).

But then comes the bottleneck. Cutting the pages, scoring the covers, assembling the pages, binding the books, and doing final trim are all very much repetitive manual labor. Fortunately, there are various machines that can automate much of the binding task.

Not too surprisingly, the more a binding machine does, the more it costs. As these machines are shop equipment, and made in short production runs, and because the market tolerates the high prices, some cost a lot more than it would seem they are worth—especially the lowest of the low-end models.

But no machine is truly expensive if it can save you time and increase your productivity. A good binding machine is going to be worth a lot if it can help you produce a better-looking product of higher quality that you can sell more easily, and sell for more money.

That is, of course *assuming that you can sell everything you print.* The whole point of book-on-demand printing is that you print only what is need. And if you're like most book-on-demand printers, you're working on a shoestring and can't afford to make many mistakes. Print what you don't need too often, and you're dead.

If you *can't* sell what you print, all a faster production setup will do is let you waste time, effort, money, and materials more efficiently. If all you can sell is three copies of your book at the monthly craft fair, you don't need a hundred-book-an-hour binding machine.

One last general note: Once again, I would urge you to try and get a look at any machine before you buy, and try and get samples of work bound on the machine. Judge the strength, flexibility, and appearance of the bind. And judge for yourself. Don't do what I say, or what the salesman says. Consider what follows a guide to your research, not the end result.

With all that cautionary material firmly in mind, let's talk binding machines.

In next section of this chapter, we'll discuss tape binders, units that don't quite fit produce conventional perfect-bound books without a little extra effort.

2. Binding Strip Machines

These binders use a strip of cloth tape with thermal adhesive on the inner, or binding, side of the tape. The tape, which is somewhat wider than the book's spine, is placed in the binding machine. The pages, with same-size front and back covers, are put on top of the tape. The binder presses the tape into the spine, and onto a half inch or so of covers by the spine edge, then heats the tape to activate the adhesive and form the bind.

These books can have a very flexible and strong bind. However, they do not have one-piece, wrap-around covers. This can be gotten around by matching the front and back cover stock to the color of the binding strip. While such books won't look exactly like perfect bound books, they can look very sharp indeed.

Until recently, spine printing was not practical on these machines. Powis Parker has come up with a solution to this problem.

Fastback Spine Tape Binding System

Fastback Model 15xs Binds books up to 1½" thick. $4,195
Fastback Model 11 Binds books up to ½" thick. $2,495
Powis Printer Spine Tape Printer. $2,495
Powis Scoring Machine, $875.
Manufacturer: Powis Parker

The Fastbacks provide strong, high-quality bindings. The Model 11 is the more basic model, with few bells and whistles. The maximum book thickness of ½ inch will not be big enough for most on-demand projects, while the 15xs (an upgrade of the previous Model 15) is far more flexible and can accommodate much larger books. (Existing Model 15 units can be upgraded to the 15xs.)

The Model 15xs has a true perfect binding mode. By using a new binding strip, called the "Perfectbind" strip, you can now use the 15xs to do true wrap-around covers. The one slight drawback is that it is a two-stage process. First the pages must be bound to the inside of the Perfectbind strip, and then the cover is bound to the outside of the Perfectbind strip in a second pass. Various types of binding strip allow for binding into various forms of covers. The lay-flat (LS) and pressure-sensitive (PS) strips make it possible to bind into covers whose spines are printed with a heat-sensitive material, such as laser toner. The pages are bound into the strip using heat, and then the combined binding strip and pages are glued into the cover using peel-and-stick adhesives.

Powis Parker also has a line of what they call "Halfback" covers that work in the same binding machines. These consist of two pieces of card stock, one of which is flat, and one of which is scored to fold, so as to form the front cover, the spine, and the first half inch or so of the back cover, as shown in the diagram.

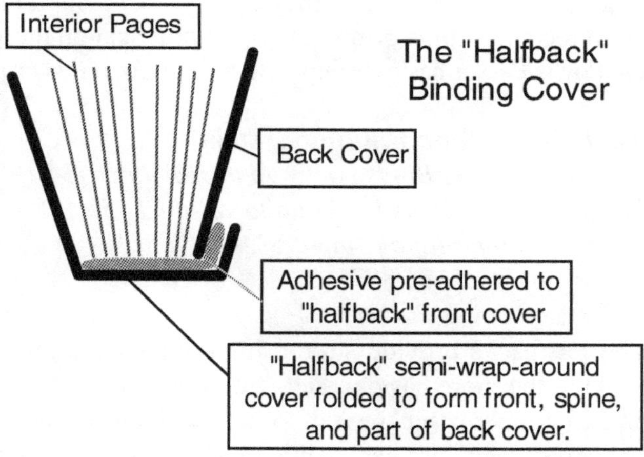

Once run through the Fastback Binder, the interior pages and back cover are bound into the front cover, producing a volume that resembles closely, though not exactly, a paperback book with a true wrap-around cover. The halfback covers can be run through an ink-jet printer (prior to binding!) in order to produce whatever cover design you like.

The Powis Printer is designed to do one job only: to print spine lettering on Powis brand binding strips. It can print black or foil spine lettering. It can be controlled from its own keypad, or else can be hooked up to a computer. The lettering closely resembles the sort of embossed type seen on fancy hardcover books. Powis Parker also sells a line of covers that match the spine tapes. Choose a spine tape in a nice cloth finish, and a cover stock that matches or complements the spine tape, and the effect can be pretty classy.

It is difficult to come up with pricing information on the supplies for the Powis Parker machines. The company itself refuses to release prices, and you have to track down a dealer. After some digging, the information I came up with is that the binding strips are sold by the carton. (Pricing is set by the distributor, and your distributor might sell at a different price and in different quantities. Still these numbers should give you a rough idea.) Depending on the width of the strip, there are either 300, 400, or 500 strips per carton, and each carton sells for $135.

The "Halfback" covers cost about 75¢ to $1.25, depending on the size of the book. The per-unit cost of the spine tapes and book covers for the "Fastback" covers is about 85¢. So, in round numbers, figure about 35¢ per bind with the Perfectbind strips (plus whatever the cover stock costs you) and about $1.00 per bound book with the Fastback or Halfback systems.

You'll need the Powis Parker Scoring Machine (discussed in Chapter Three) or a similar unit for some binding procedures.

The Fastback System and its variants produce excellent binds that have very good lay-flat characteristics. The books look good too. While the 15xs can produce true perfect-bound wrap-around covers, it involves a somewhat cumbersome procedure that requires two binding operations—one to put the pages together, and one to put the bound pages in the cover. Then there is the issue of price.

The Powis Parker hardware will cost you upwards of $7,300 if you want to do inch-thick book with spine lettering *and* cover scoring—and it can cost you about $1.00 a book to do your binding. Other machines in the same price range have a bind-cost-per-book of about one to five cents, plus whatever the covers cost you.

It's a good system, but at those purchase and per-bind prices, I would not consider it the best possible choice for a book-on-demand printer. These units are really for offices that need to do 200 annual reports, or put together snazzy promotional booklets in a rush, or bind legal documents. They can be used for book-on-demand as well, and will produce good results—but it'll cost you. However, if you were doing other projects wherein the Fastback made sense, and wanted to do book-on-demand as well, this binding system might well be the way to go.

Accubind Document Binding System

Maximum Binding Width: 1 inch
approximately $4,500
Manufacturer: Standard Duplicating

The AccuBind Document Binding System is a tape binder, similar to the Fastback as described above. It allows for lay–flat binding. It's a less elaborate system than the Fastback. It does not appear to have any facility for spine printing. It does not allow for wrap-around covers.

However, there is no particular reason why books bound with a spine tape machine could not be glued into case bindings (the "covers" of the paperback book in effect becoming the endpapers) to produce case-bound books with very flexible adhesive spines. Such books would look very much indeed like conventional hardcovers. It's worth considering. (See the Fastbind entry below for their hardcover binding system.)

3. Other Small Binding Machines

These are machines that don't fit into any one category, but are worthy of mention. They use somewhat unusual ways to produce standard perfect-bound paperback books. Let's start off with a look at a machine I've never actually seen, despite years of trying.

BQ-P6

"Hybrid" Binder
Maximum Binding width: 1 inch
Weight: 84 pounds
Manufacturer: Standard Finishing Systems
Approximate Price: list $4,995, but may be discounted to about $3,600

The BQ-P6 (now there's a catchy name) is a table-top binder that is supposed to do spine tape, pad, and true wrap-around cover perfect binding. (Pad binding is essentially perfect binding without a cover. Adhesive is painted over one edge of a stack of paper. Result: a pad of paper wherein you can peel off the sheets from the top, one at a time.) The BQ-P6 is small: 27 by 18 by 11 inches, and it is credited with the ability to do 140 perfect binding an hour. All I've seen of it at this point is the product brochure, and that brochure says nothing at all about its mode of operation. If the specs are to be believed, and if the binds are at least reasonably good, this machine could be a very good choice for small-scale book-on-demand work. However, my information is pretty sketchy. I have never seen the actual machine demonstrated at any of the trade shows I've attended, and every request for more detailed information has resulted in my getting yet another copy of the same brochure.

As we'll see a bit later, this unit sells for something close to the same price as the Bind-Fast 5, also made by Standard. I spoke to a sales rep and asked why they sold two machines in direct competition with each other at a single price point. He explained that the BQ-P6 is a lighter-duty unit, intended for something more like an office environment, where it would be called on to do a little bit of everything, rather than for the specialized heavier-duty work the Bind-Fast is meant to do. Maybe that explanation is for real, or maybe it's double-talk to cover up the fact that two machines at one price makes little sense. At $3,600, I'd at least consider the BQ-P6, if I could learn more about it. If the BQ-P6 was selling for $4995, I'd go with the Bind-Fast. Do your homework.

Fastbind Binding System

Manufacturer: Maping (Finland)
U.S. Distributors: ExactBind Corporation
Maximum spine length: varies11.8 inches.
Various Models:
***Secura**: Price: $2,600 (options could make it $3,000)*
Maximum binding width: 1.75 inches Weight: 77 pounds.
***290**: Price $5,400 (options could make it $5,900) Maximum binding width 2.2 inches.Weight: 92 pounds. Note: The Model 290 is the same unit as the European Semimatic R2, but modified to run on 110 volt current.*
***321NR**: Price: $15,000 Maximum Spine Width: 1.5 inches Weight: 155 pounds.*
***H530** Hard Cover Maker: $1,200*

These machines represent a very interesting new arrival, and operate in a somewhat unusual manner compared to other binders. See the diagram on page 134.

The heart of the system is a clamping system wherein the user loads the cover and the book block in together and then clamps them to each other, so they are kept precisely in position relative to each other.

The combined book-block and cover are loaded in spine-edge down. with the back cover held in with the pages, and the spine and front of the cover outside the clamp. The spine and front cover are then folded back out of the way, exposing the spine-edge of the book pages. Then the whole clamp assembly is swiveled about so the spine edge is facing up. A roughening head (excluded on the Secura and optional on the 290) is run over the spine of the book block, from one end to the other. Then a gluing head is run over the roughened spine edge of the book block, applying a layer of hot-melt adhesive to the entire length of the book. Because the spine is facing up during adhesive application, the glue will tend to seep down just a bit into the book block, strengthening the final bind. The spine and front of the cover are then pulled around into position over the glued spine of the book block. Finally, the clamp assembly is swung back around to the spine-down position, and a nipper clamp presses in, forming the book cover around the inside pages. The bound book is then removed from the machine for trimming.

Maping lists a "saw-brush" on their web site, for use in manual spine roughening. That might be a handy item for the Secura, or for any binder system that didn't include spine roughening.

The higher-end Fastbind machines include a built-in trimmer for cutting away any excess length of front cover stock. However, this didn't seem particularly useful to me. Unless all your measurements are always exactly correct, and all of your pages are always perfectly jogged into place, and nothing is ever slightly misaligned, the books will require a trimming run through a paper cutter anyway.

There are several machines in the Fastbind line beyond those listed, with the cheaper models using completely manual operation. The most expensive machine is highly automated, and the mid-range units are partly manual and partly automatic. The U.S. distributors, *Exactbind* currently lists the model described above, but it couldn't hurt to check their website or Maping's, and if one of the other units seemed right for you, you could query the U.S. distributor.

As regards cover preparation: You *might* be able to get away with no scoring with very light cover stocks. However, the Fastbind system really relies on very sharp and exact creasing or scoring of the covers. Fastbind books actually only require one crease on the cover, on one side of the spine before it goes into the machine. The second spine-edge score is formed during the binding process. The binders are designed to put some glue on the front and back pages of the book-block, so that a narrow strip of the front and back covers are glued to the book block. The nipper clamp squeezes this glue in tight as the cover is formed around the book-block. A thin line of glue on the edges of the front and back pages is a good thing, as it results in a stronger bind and takes some of the stress off the spine when the book is opened. But it also means that the cover really should be creased

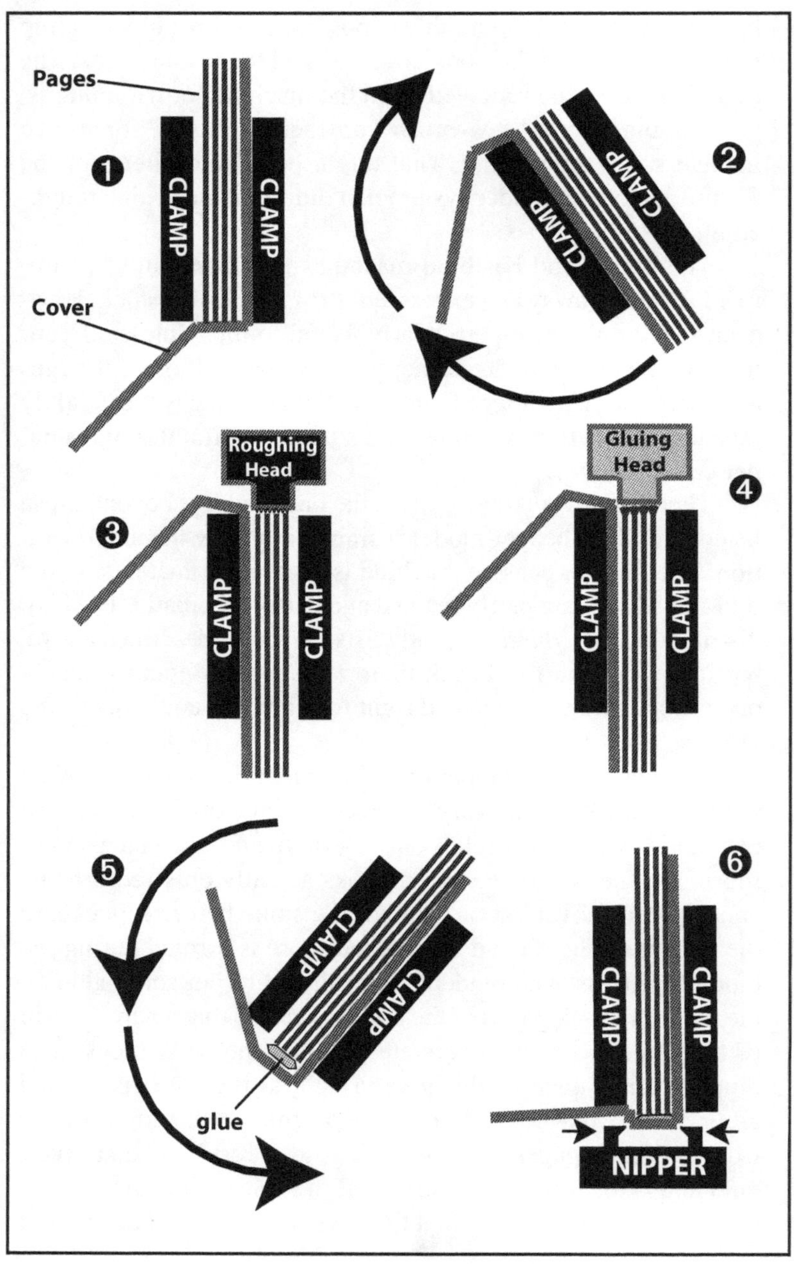
Pages
Cover
❶
CLAMP
CLAMP
❷
CLAMP
CLAMP
Roughing Head
❸
CLAMP
CLAMP
Gluing Head
❹
CLAMP
CLAMP
❺
CLAMP
CLAMP
glue
❻
CLAMP
CLAMP
NIPPER

The Fastbind System

1. The pages and cover are clamped in place, spine edge down. The rear cover is clamped alongside the pages, with the creased spine and front cover outside the clamp.
2. The clamp, holding the pages and cover, is rotated 180 degrees so the spine edge is up. The cover's spine and front are folded out of the way.
3. The roughing head is passed over the spine-edge of the pages, moving from one end of the book to the other. Because roughened pages have more surface area, this exposes more surface for the glue to touch, improving the bind.
4. The gluing head is passed over the spine-edge, leaving a layer of hot-melt glue on the roughened edge of the spine.
5. The spine of the cover is folded over the spine edge of the book, and the entire book is rotated back to the spine-down position.
6. The nipper clamp presses in on the spine, forming the cover firmly around the pages. The book is then removed from the binder for trimming.

a short distance away from the spine on either side to allow the covers to hinge open smoothly.

There's also the question of aesthetics. A paperback book with a very sharp, square-off spine edge on front and back looks good. A book with a squared-off front spine edge and a rounded, less square rear spine will look odd. Even if the Fastbind system only requires one crease, the books will look better with all the creases there.

In short, your covers should really have four scores or creases—one on either edge of the spine, and one more on the front and back covers, about a quarter inch away from the spine. Because the spine needs to be exactly the width of the book-block, you'll need high precision on these creases. Some sort of precise scoring table probably should be budgeted into the purchase when considering one of these units. ExactBind sells a Fastbind manual cover creaser for $600 and an electric one for $1,700.

It should also be borne in mind that the cover-scoring is one more bit of hand-labor. Additional labor per book is no big deal on very short print runs, but if you're doing 200 books, that means you're hand-creasing 200 covers.

See Chapter Three of this book for additional discussion of scoring and scoring machines.

The Fastbind system can also be used to bind hardcover books, and there is also a separate machine, the H530 Hard Cover Maker, for forming the casing of a hardcover book. The covers can be of any material, but of course the cover stock will either have to be very robust to start with, or else well-laminated, in order to hold up to use. Creating a hardcover book with Fastbind will require some handwork, and you'll need to use the special self-adhesive endpapers and other supplies made for the system.

I saw a live demonstration of the Fastbind system at Book Expo America in 2002. I was very impressed with the quality of the bind on the paperback, though I was bit concerned about the need for finicky and precise cover-scoring.

I would rate the quality of the paperback book I saw produced as excellent, but the hardcover book was only acceptable. In fairness, the hardcover was produced by a operator who was rushing a bit toward the end of the last day of what had been a very long trade show. While I can only report on what I saw, I expect a more skilled and careful operator could produce a much better hardcover. The main thing is I did see someone produce a hardcover book by hand in five minutes. That's got to count for something—in fact, for quite a bit. Exactbind claims a unit cost under $3.00 per book to make up the hard cover.

To sum up, the Fastbind system looks to be an innovative and economical way to make books. My two caveats: (1) Factor in the cost of the creaser, and the additional hand-labor required for creasing. (2) Insist on getting samples of hardcover books—and examining them closely—if you're interested in doing hardcovers yourself. All that being said, this looks like a pretty nifty system.

The machines we have just discussed in this chapter all work in slightly different ways. The machines in the next section are all basically variations on a single theme. They have different features, but all are based on the same principle, and work in pretty much the same way. Therefore, we'll start the next chapter with a quick discussion of how they're all the same, before we find out how they're different from each other.

Chapter Eight
Production Binding Machines and Specialty Large-Scale Machines

1. General Operating Principles of Most Small Commercial Binding Machines

The rest of the binding machines we'll look at can be considered as a group. All these binders work in just about the same way, with each machine a bit more elaborate than the last.

It's worth noting that virtually every paperback book—and many of the hardcovers—on sale in your local bookstore are bound together by machines that work in exactly the same way as the ones we're about to discuss. In fact, more than likely, you own books that were bound on one of the machines listed below. Small press titles and/or books produced in shorter run might well be bound on a BQ-140 or the like. However, most paperbacks are produced on fully automated binders designed to pound out four hundred copies an hour or more. Those machines work in essentially the same way as the machines below. They just do the work faster, and more automatically.

So here's how 99 percent of all adhesive binders work, in very general terms.

The book pages are dropped into a clamp, spine edge down. The unscored cover is placed, face down, on a horizontal bed, sometimes called a nipper table, on the opposite end of the binder. Various paper positioners see to it that the spine of the binding is lined up with the book pages and held in place.

The page-clamp is on a movable stage. An electrical drive system moves the clamping stage toward the cover, first passing over a roughing, notching, or milling unit (these are omitted in the cheaper units, and I'll explain the terms in a minute). The page-clamp then carries the pages over a glue reservoir. Usually this is a wheel, spinning in the direction opposite to the clamping station's travel. The wheel picks up glue (either hot-melt or cold depending on the unit) from a pot below the binding stage and paints it onto the spine.

The page-clamp, carrying the now glue-painted pages, moves past the glue applicator and into position over the cover. An automatic or manual system presses the cover onto the pages by means of "nipping" clamps onto either side of the cover spine. This creases the cover sharply and presses it into shape. The book is removed from the binding machine and set to one side, where the glue is given time to dry or cool.

Roughening or *roughing* is exactly what it sounds like, a way to roughen the spine edge. The rougher the surface, the more surface area it will have, and the better it will hold glue, and the stronger the bind will be. Roughing is done with small rotating blades that chop into the spine as it passes.

Notching is also just what it sounds like: notches are cut into the spine. These likewise provide more surface area for the glue to get at, and make for a stronger bind.

Milling is the process of actually cutting away a portion of the spine edge of the pages. This is necessary for signature-printed pages, because the final fold has left all the sheets of each signature nested one inside the other. Milling chops away the folds, exposing the inner pages of the signature to the glue. Roughing and notching are good things, but, for the normal book-on-demand book, milling isn't necessary, as the "signatures" are cut apart (for example, the letter-sized sheets cut into sets of 8½ by 5½ inch pages) before they reach the binder.

Only the more expensive binding machines include roughening, notching, or milling units. Without these steps, about the only way to make for a stronger bind is to use more adhesive, and the more adhesive you use, the stiffer the binding, and the harder and more awkward it is to hold the book open.

You could probably rig up some sort of separate roughing or milling machine. All you'd need some way to clamp the pages while you ran them over a sander or a power-drill with the right sort of sanding or grinding attachment. *Maping*, the company

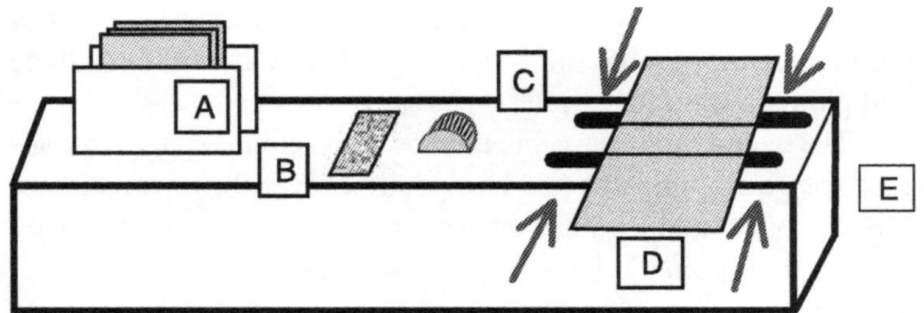

Basic Operation of A Small Binding Machine

The book pages are placed in a clamp [A] that moves over a notching, milling, or roughing head [B] (this step omitted in lower-end models), over a glue applicator [C], and into position over the cover [D]. Nipping clamps [E, indicated by four arrows] press in on either side of the cover, folding and creasing it into position around the glued book-block. The book is removed from the clamp and set to one side until the glue is completely dried or cooled.

that makes the Fastbind system, sells a "saw brush" intended for roughening covers. However, most of the binders that don't do roughening or notching still produce good strong binds.

Really expensive units, way out of range for any book-on-demand printer, also include automatic cover scoring. Though most of these machines we'll look at claim that no scoring is needed, what they mean is that it is not needed if your cover stock isn't too heavy.

In other words, not scoring the covers on the units we're about to examine is strictly on a when-you-can-get-away-with-it basis. If your cover stock is too heavy, and/or if the grain of the cover stock is running the wrong way, you will have to do manual scoring, and do it with sufficient precision for the scoring to line up properly with the book pages. However, if you pick a cover stock that doesn't need scoring, you can avoid this step.

These true thermal perfect binding machines will likely be easier on your covers than the office-type binding systems we discussed earlier. The office binders heat the cover from the outside, and rely on the heat moving through the cover to transmit the heat to the glue *inside* the cover. This means there has to be a lot of heat passing through the outside of the cover spine, and that heat can melt the laser toner right off the cover stock.

True thermal perfect binders melt the glue and apply it to the pages themselves. The covers are pressed into place so that the hot glue hits the *inside* of the cover, and the glue is already starting to cool by the time the pages and the cover meet. Thus, there is a lot less heat applied to the outside of the cover, and therefore there is less potential for damage.

A production thermal binder might or might not do damage to laminated covers, depending on the laminate material. It could cause one hell of a mess if the binding machine melted the laminate. Obviously, it would be the laminate on the inside of the cover's spine that would come in direct contact with hot glue and thus would be most at risk. I have seen books with fully laminated covers and thermal binds that were acceptable, though not outstanding. *GBC* makes a special "glueable, stampable" laminate that's designed to deal with hot glue.

All of the binding machines discussed below require the full attention of an operator to feed covers and stacks of pages, work the controls, and remove the bound books. Just about all of them claim operating rates of over a hundred books an hour, and some claim rates of two or three books a minute. In my opinion, that rate of production on any of these machines would require an operator on some sort of illegal substance.

From what I've been able to see, all of the claimed production rates are unrealistically high and therefore misleading, which is why I have not included them in the specification lists. The operator will be the limiting factor. The question is: how many books can *you* deal with per hour? My hunch would be that one book a minute, maybe two, tops, would be a realistic production rate for any of these machines in the hands of a part-timer operator.

Pretty much all of the brochures for these machines urge you to use only their adhesive, specially formulated for their machine. In the section on office thermal binders, I argued that it was silly to assume that Brand X binding covers would not work in a Brand Y machine. However, things are different with the true production binding machines. When it comes to true production binders, the risks of using off-brand adhesives are higher, and the potential benefits are lower.

These glues are designed to melt to a certain viscosity at a certain temperature, and the machine is built to deal with that thickness of glue and that degree of heating. As opposed to the adhesives in the thermal binders, which just stay in one place and melt, these glues are also meant to be moved through mechanical systems. Typically, a thermal adhesive for book binding is supposed to be taken up by a rotating glue applicator wheel, over which the book block moves, spine down. The glue is in essence painted onto the spine of the pages, and has to stay on the spine without dripping off as the book block is carried along. The glue must retain enough heat to still be liquid when the book block gets to the waiting cover. The glue needs to be a liquid as the cover is nipped into place, but then cool back to a solid within a few seconds, when the book is removed from the machine.

Obviously, a glue that can do all that must have quite specific characteristics for heat retention, viscosity, melting point, and so on. A given machine might be designed for an adhesive that is the consistency of room-temperature honey at 335 degrees. Dump in an adhesive that's the consistency of cold porridge at that temperature and the glue applicator might jam.

The makers of the office thermal binder units charge jacked-up premium prices for their "custom" glues and covers. The commercial systems price their glues as what they are: commodities. If the recommended glue costs you 3¢ a bind, how much are you going to save by using a glue that costs half as much, but might not be formulated properly for your binding machine?

Therefore, assume that you're going to be pretty much married to whatever adhesive your system uses. Get reliable price quotes on the adhesive for any machine you might consider, and make sure you have a steady, reliable supplier for the stuff—

better still, multiple suppliers. This does not necessarily mean you have to buy adhesive from the binding machine's manufacturer. It might well be that company X makes an adhesive that's "specially formulated for, and meets all manufacturer's specificiations for" company Y's machine. But be careful.

Just by the way, the adhesives for these machines are sold by the pound or the kilo or the gallon, and used by the tenth-of-an-ounce by the binder. You might have to buy a case of adhesive for $150, but that case might well be something close to a lifetime supply for a low-volume producer

Then there is the issue of the strength and flexibility of the various adhesives. As we have seen, you want the glue strong enough to hold the book together, but you also want a book that you don't need to pry open with a crowbar. As flexibility varies not just with adhesive composition and adhesive thickness, but also with page size and spine thickness, and paper grain orientation and other things, try and get samples of books of various thicknesses bound with the binder you're interested in. Obviously, try and see a sample of a book as similar as possible to the size you most want to print.

Also make sure that you can get your machine serviced, or satisfy yourself that you can service it yourself. These are rugged pieces of hardware, but if your binding machine dies, you're out of business. If you buy a used machine, make sure there is a local dealer who is willing to repair your machine even if you purchased it elsewhere.

For the most part, these binders are made by companies that also make machines with tens or hundreds of times the capacity of these machines. Even the priciest of these binders is small potatoes in the printing world. Selling to *you* is therefore small potatoes. Be prepared for salesmen who'd rather be selling a $250,000 binder to Conglomco Enterprises.

Finally, it should also be noted that just about all these binding machines can do *padding*, which is the process of making pads of paper such that one sheet at a time can be peeled off from the top of the stack. Basically, padding is binding without a cover and with less adhesive, making for a deliberately weaker bind. As you can also do padding with two pieces of wood, two clamps, a paint brush and a $10 bottle of padding compound, this feature probably isn't a good reason to buy a machine costing several thousand dollars.

It is generally possible to locate used binding machines through various print-shop want-ads, both on paper and on the Internet. Whether or not that's a good idea, I don't know. With some things, like anvils, it is safe to buy used, and with others, like brake pads, it is risky. Small binding machines seem to fall somewhere in between. For the most part, they seem to be built to last, but they can see very hard use. I bought a banged-up Bind-O-Mat 200 (a binding machine that's no longer made and won't be missed) and got it more or less up and running, but suffice to say the previous owners hadn't treated it with great respect, and I gave up on it. I then bought a used, but well-maintained, Bind-Fast 5, which works very well indeed. You have to shop carefully.

These machines aren't meant for the consumer market, and thus it's hard to find out about them via consumer channels. It was a real struggle to gather information on these machines. My initial information on these machines came mostly from manufacturer's brochures. At the time I did the first draft of this book, I had laid eyes on a small commercial binding machine exactly once. It was a Bind-Fast 5, and I saw it at two a.m. at the back of a Kinko's copy shop when I stopped in there after a long poker game. I didn't get much chance to look it over too carefully. I've done better since, but the information in what follows is still second-hand to a certain extent.

I have omitted discussion of machines that are way too big and fast for purposes of book-on-demand. Even the selection of machines I do discuss go beyond what most book-on-demand operations could ever need.

In the following sections, we'll discuss the specs of the various machines. The prices I list on each unit are approximate. Remember the manufacturers can introduce and discontinue new models, and raise and lower prices, whenever they like. Some units are imported, and prices can yo-yo as the dollar and the pound and the yen (and now the euro, I suppose) go up and down.

Which brings me to another point, one that is applicable not only to binding machines, but all the binding hardware you purchase: do your homework regarding supplies and the viability of the manufacturer (and, in the case of an imported machine, of the U.S. distributor) before committing to a machine intended for production work.

Be as sure as you can be that the machine, the spare parts, the supplies, the manufacturer, and the company that sell you production equipment will all be there in the morning. In principle, this is as important for U.S. companies as foreign ones, but importing binding machinery is apparently a volatile business. Five true binding machines that were listed in the first editions of this book are not listed in this edition, because they are no longer available. One was a decades-old domestic design and a piece of junk (the aforementioned Bind-O-Mat). The other four were imports that aren't being imported anymore.

It's no big deal to the average TV buyer if Hitachi leaves the U.S. market once he or she has purchased a TV. The TV won't need a constant supply of Hitachi-brand electrons to keep working. Not so with a binding machine that needs specific, even proprietary, supplies in order to operate. If the manufacturer stops making the adhesive, and you can't find a suitable replacement, you're stuck. If the machines breaks down, and parts are no longer available, you're even more stuck. If you're dependent on a single source for raw materials, you'd better make sure that source is going to be there in the morning.

You'd also better make sure the machines themselves will still be around. The last time I did a round-up of binding machines for on-demand printing, Planax had three table-top thermal-binding machines in the U.S. market. All three are now gone from the U.S. market, and possibly altogether, and Planax has gone to all cold-glue binders. I don't know what you'd do about spare parts if you bought the last Planax Autotherm before they pulled the product. In similar fashion, as discussed in a previous chapter, Unibind dropped the Penta-Bind line shortly after I bought it. The sole source of official Penta-Bind binding sleeves that I know of is in the top of my closet. In short, make sure replacement parts, or, even more importantly, binding adhesive, will be available, and available locally. You don't want to deal with supplies that have to be shipped from Germany or Finland or, far worse, some third-world Trashcanistan every time you place an order.

There's a lot of relabeling that goes on in the business world. It wouldn't surprise me a bit if many of the allegedly domestic models turned out to be foreign-made and merely distributed by a U.S. company.

Pricing for binders in general seems to be fairly negotiable. I have list prices shown below, but I have found significantly different prices on the same model. But bear in mind that shipping and set-up costs are going to be significant issues with any of these machines.

But enough general commentary. Let's get to the binders.

2. Table-top Binding Machines

These are the smaller machines, that can indeed fit on a tabletop. They are more limited than their big brothers, but generally in ways that really won't matter to the home book-binder. They will get the job done.

Bind-Fast 5

Manufacturer: Standard Finishing Systems
Approximate Price: $5,300, street-price quote $4,200 Often available used for $1,500-$4,500
Maximum binding width: 1¾ inches
Maximum spine length: 17½ inches
Thermal binder, binding temperature 340° F, weight 100 pounds.

This tabletop machine, as best I can see, is the Volkswagen Bug of binding machines. It's everywhere, everyone knows about it, it's simple, it leaves off features you might want but probably don't need, and it will get you where you want to go.

The Bind-Fast operates as follows: an electric carriage moves the binding clamp over the glue applicator and into position over the cover. The nipper table has a manual lever-operated nipper clamp. The Bind-Fast 5's nipper requires a manual adjustment, but that adjustment is fast and easy. The Bind-Fast 5 does not do roughing, notching, or milling. However, it does produce a strong and reliable bind.

Mote that you need the optional nipper table to do perfect binding. (The $5,300 price above includes the nipper required for perfect binding.) Without the nipper, you'll only be able to do padding and report binding. (Report binding means putting the pages between front-and-back covers: that is, binding with no cover on the spine. Usually there is some sort of decorative spine tape applied to hide the raw adhesive).

I am not entirely clear on one issue concerning model numbers: Either the model that can take the nipper table is the 5c, or else *adding* the nipper turns a mere 5 into a 5c. Check into this, because according to one source, the nipper table won't fit onto some early versions of the Bind-Fast. I bought a used 5c, and I like it a lot. I used it to bind the home-printed version of the book you're reading. On my first try, it did a better bind than I had ever gotten before.

The Bind-Fast 5 is probably the best binding machine for the average on-demand bookbinder. It is a serious, functional, easy-to-operate machine, though it does have some limitations, and omits some features, such as roughening or milling.

Also as with the Bug, it seems as if there are often used Bind-Fasts floating around. When cruising various out-of-the-way bits of the Internet, and various specialized trade magazines, I have spotted a number of used Bind-Fasts for sale, at times for well under $3000, or even under $1000. What sort of condition such machines are in, I have no idea. Shop carefully. Make sure any used unit you buy is equipped for perfect binding, or at the very least that the nipper table will fit on it.

DB-250

Manufacturer: Duplo
Distributor: Duplo USA
Approximate Price: $7,000, probably negotiable down to $6,000
Maximum binding width: 1.6 inches
Maximum spine length: 15 $^{3}/_{4}$ inches
Thermal binder, binding temperature approximately 360° F., weight 255 pounds.

The DB-250 is more automatic than the Bind-Fast. Unlike the Bind-Fast, the Duplo does roughening and notching, thus improving the strength of the bind by exposing more surface area to the glue.

This unit is likewise supposed to be a table-top unit, but if "table-top" can apply to a machine that's over 250 pounds, it's clear that the term is pretty flexible. The DB-250 includes a "unique" roughing/notching system that, at least according to the brochure, "eliminates paper dust for cleaner operation."

This unit, and most of the others listed below, are two-pass machines. The book pages are loaded into the clamp on the right side of the machine, and the binder is started up. The clamp travels right to left over the rougher/notcher. Once the pages are in place on the left side of the machine, the book cover is loaded into a nipper table on the right side of the unit. A sensor prevents the page clamp from traveling back until the cover is in place. (This sensor can be shut off for padding work.) With the cover loaded, the page clamp carries the pages left to right, back over the rougher again, then over the glue pot, and then into place over the cover. (A mechanism keeps the glue off the pages on the outward pass.) The automatic nipper table then forms the cover around the book pages, and the bound book is removed from the binder.

Other features include a built-in counter to clock how many books you've done, an electronic rather than mechanical control panel, and an emergency reset button so big and red and right in the middle it's bound to get slapped by accident once in a while. In short, the DB-250 has more features than the Bind-Fast—not the least of which is that roughing and notching head. Its claimed specs and features rival or match all the more expensive machines in this summary.

The DB-250 is still a relatively new machine, but it has now been in the U.S. market for a few years. The last time I saw it at a trade show, I was assured that the 250 had gotten so many upgrades since its introduction that it was practically a new machine. That's as may be, but the basic specs have remained the same—though three difference sources listed the maximum binding width three different way —as 1.4 inches, 1.5, and 1.6. I don't know if that's a rounding error, hype, or truly increased capacity.

This is an intriguing machine. I saw it demonstrated and it did a nice job—though the sales rep doing the demonstrating had some trouble with the super-sophisticated controls.

Duplo USA is the California-based American subsidiary of Duplo, a Japanese manufacturer of binding equipment. (The DB-250 is actually made in Taiwan.) As with just about everyone else in this round-up, Duplo mostly makes bigger and heavier-duty units, with the table-top unit very much the junior partner.

The DB-250 is a relatively new model, in a business where most machines stay in production for years or decades. I have heard a few mutterings from Duplo's competitors about reliability and support, but I don't know how much stock to put in these reports. As with all of these machines, research issues of warranty, product support, repairs, and parts before reaching for the check book.

Bourg Perfect Binder BB1000/BB1001

Manufacturer: C.P. Bourg
Price: $7,700
Maximum binding length: 14.33 inches.
Maximum binding thickness: 1½ inches. Weight: 265 pounds.
Glue temperature: 239⁰ to 392⁰ F.

This unit does notching and roughening, and includes a paper waste removal system. This binder looks to be a competitor to the Duplo 250. It appears to be a solid, powerful machine suitable for higher-end book-on-demand work, though without some of the non-essential bells and whistles on the Duplo.

Some of CP Bourg's literature discussed the BB1000, while other information was about the BB1001. I am not clear whether the different model numbers mean the 1001 is an upgrade, or whether the numbers are meant to distinguish two sub-models; for example, one unit is for U.S. electricity and one for European.

The wide range of glue temperatures suggests that it might take a range of adhesives for differing applications, or else that the glue formulated for it had different characteristics at differing temperatures.

This unit is very much at the low-end compared to most of Bourg's products. It would not surprise me if this unit turned out to be made by someone else and marketed by Bourg in order to fill a hole in their product line. There's nothing wrong with that, but it would be worth checking out insofar as service and supplies are concerned.

Note: the info sheet claims the unit does "scoring," but this does not mean it scores the covers. It would appear to be a reference to spine-notching by another name. The same info sheet specifically states that with thick cover stock "prescoring of covers may be required."

3. Floor-Model Binding Machines

Most floor-model machines are probably much too big and expensive for any small-scale book-on-demand operation. I discuss them here to give some idea of what the next step up is, and also because we'll be discussing large-scale book-on-demand later in the book, and machines such as these are what the big shops have. Hire someone else to bind your books for you, and they will likely use a larger floor-model unit machine such as those listed below. Even if you can't get within ten thousand bucks of affording one of these machines, it might well be useful to be somewhat familiar with the specs and capabilities of a larger binding machine.

Once again, the books-per-hour claims are so much vastly higher than what a normal human could achieve that I see no point in listing them. From this point on up in price, part of what you are buying is not just speed, or bind-quality, but ruggedness. The machines we've seen so far are for occasional use, or for a hour's worth of work a day. Pound them too hard and they'll just plain wear out. The machines below are expected to be on and running all day every day.

850 Series

Manufacturer: Rosback Company
Various Models: 850H (manual clamp & nipper) $8,900, 850M (manual clamp, auto nipper) $12,500, 850ME electric clamp, auto nipper) $14,440, 850MEC, electric clamp, milling, auto nipper, fume/chip exhaust) $17,450
Maximum spine width: 2 inches
Maximum book length: 15 inches
Thermal binder, binding temperature unspecified, weight 225 pounds.

The 850 series brings an interesting approach to short-run binders: offering the buyer a basic model, as well as variants with more and more features. As can be seen from the prices listed above, you can come close to doubling the price of the basic unit by adding features. The basic 850 has a notching head, and a self-adjusting nipper head—in other words, it adjusts itself to the spine width. The higher-end models add more and more features, but Rosback claims the same production rate for all the machines. What you're getting with the higher priced models is more automatic (and therefore probably more precise) operation. The MEC model adds a milling head, which means it can deal with signatured pages. The units are not upgradeable. According to the Rosback rep I spoke with, it is not, for example, possible to add the auto-nipper to the basic model and turn it into an 850M. By offering a range of options, Rosback is offering one machine design that competes against models across a range of price points.

BQ-140

Manufacturer: Standard Finishing Systems
Approximate Price: $12,000
Maximum binding width: 1.2 inches
Maximum spine length: 15.7 inches
Thermal binder, binding temperature unspecified, weight 242 pounds

The BQ-140 has a notching head, a vacuum system to clear out the dust produced by milling, and an electrically operated automatic self-adjusting nipper clamp. The BQ-140 comes on casters so it can be rolled out of the way when not in use.

Once again, the BQ-140 has a very high claimed production rate. While the production rate will no doubt be higher on this machine than on the smaller machines, a lot of what the extra money buys you is a *better* bind, not a faster one.

There are a lot of BQ-140s out there, which means there is a much better chance that you can find parts, supplies, and repair service available as compared to less common machines.

Oddly enough, none of the spec sheets I have on the BQ-140 actually said anything about whether it is a cold-glue or thermal binder. I had to talk to a salesman, and watch a pretty dull promotional videotape, to confirm that it is a thermal binder, without a cold-glue option.

One potential drawback to this machine is that it has one of the narrowest maximum binding widths. It will only bind books up to 1.2 inches thick.

Bourg Binder BB2000

Manufacturer: C.P. Bourg
Approximate Price: $15,000
Maximum binding length: 17.7 inches.
Maximum binding thickness: 2.35 inches. Weight: 540 pounds.
Glue temperature: 284^0 to 392^0 F.

This unit's claim to fame is the maximum binding width of 2.35 inches—very close to twice the spec on the BQ-140. It includes a spine notcher and dust removing vacuum system, and boasts a safety shield to protect the operator, tool-free set-up, and "easy-to-use" microprocessor controls. (Of course, none of the units here require tools for normal set-up, and they all have some sort of electronic controls, but they have to put *something* on the sales sheet.) As with the BB1000/1001, the BB2000 lists a range of glue temperatures. I am not entirely clear what the purpose of this is, unless they offer a range of glues for different tasks.

This unit is eleven inches longer than the BQ-140, and weighs 300 pounds more. This is very much a heavy-duty piece of equipment, one that's intended to take a lot of day-in, day-out use.

BaumBinder 300

see listing under cold-glue binding.

4. Commercial Cold-Glue Binding and Lay-Flat Binding Machines

While there are plenty of binding machines pricier than the BQ-140 and BB2000, my guess is that even these units are a lot more machine than anyone reading this is ever likely to need. All of the manufacturers listed above, as well as companies like Sulby and Muller-Martini and a number of others, make a wide variety of high-end machines, but even someone as enthused about hardware as myself can't see any justification for reporting on such machines in this guide.

You can spend $250,000 or more on a binder with no trouble at all. At about the $25,000 mark, the machines tend to get more and more automated, and more and more geared to longer production runs, with machines designed to take in stock from other machines, rather than human operators, and machines designed to feed bound books straight into an automatic paper-cutter for the final trim. However, Planax, God bless them, does sell the cool-looking but incredibly overpriced Perfect Binder FII, which has none of these features for a mere $25,000. We'll come to that unit in a minute, after a brief detour.

BaumBinder 300
Hot-Melt/Cold Glue Binder
Manufacturer: Baum Folder
Approximate Price: $17,000
Maximum binding width: 2 inches
Maximum spine length: 17 inches
Glue temperature in hot-melt mode: unstated.
Weight approximately 484 pounds

This unit is the only binder remotely close to being affordable for the on-demand binder that can do both cold-glue and hot-glue binding, meaning that it offers a high degree of flexibility. I saw it demonstrated in the hot-glue mode, and came away with a sample book that had razor-sharp corners on the spine fold, and a very solid yet flexible bind. It has a milling head that will cut away up to $^{3}/_{32}$nds of an inch (2 mm) of paper as well as a notching head. The milling head means that this unit can bind together pages that arrive at the binder folded into signatures.

The switch from cold to hot glue is achieved by swapping out the glue pots. While the brochures promise this procedure is simple, I doubt I'd want to be switching back and forth every day. For most binding jobs, hot-melt is probably the better choice. But for those jobs that require a cold-glue bind, for whatever reason, it is nice to have the second arrow in your quiver.

There are various clues and indications that tell me this is likely an imported unit added to Baum's line of paper-handling equipment in order to round out their offerings. Baum does not seem to offer any other binding machines, whereas all the other manufacturers have a full range of binders (of which only the low-end are listed in this book). That being said, this unit offers a lot of flexibility and quality for the money.

Perfect Binder FII

Manufacturer: Planax
U.S. Distributor: Import-Graphics
Approximate Price: $25,000
Maximum binding width: 1.375 inches
Maximum spine length: 15 inches
Cold Glue Binder, weight approximately 606 pounds

The FII is more or less comparable to the BQ-140 in terms of specs, features, and operation, (but with a 1.375 inch maximum binding width), except that is a cold-glue binder. It just costs twice as much. Why it should cost an extra $13,000 to squirt cold glue instead of hot, I don't understand. The price difference comes near to being the complete price for the BaumBinder 300.

However, there is at least a certain degree of value for money in this machine for certain applications, at least potentially. Its claim to fame is that its cold adhesive is flexible enough to allow for a lay-flat binding.

The average thermal-bound paperback will not stay open on its own. All other things being equal, a book that stays open is to be preferred over one that snaps shut. Certain types of books more or less have to have lay-flat bindings: sheet music, cook books, picture books, and so on. Supposedly, the cold-glue adhesive used by the FII is strong enough, and yet flexible enough, to do lay-flat bindings.

Otabind/RepKover

Planax also sell a gadget call the RKM 200, part of what they call the Repkover system, a variation on a patented process called Otabind. The RepKover/Otabind process is likewise supposed to insure lay-flat binding. You use it prior to normal binding operations. It lays a piece of special cloth tape down the spine of a book cover and scores the cover. It glues the tape, not to the cover's spine, but to the inner edges of the front and back covers. The book pages are then bound into these special covers in the normal way in a normal perfect binder.

The easiest way to understand why this is a good idea is to look at the binding of a good-quality hardcover book. Hold the book spine-up and open it. If you peek down the spine between the cloth cover and the pages, you will see nothing but air and daylight: the book pages are not attached to the spine at all. Instead, the endpapers are glued to the pages, and to the front and back covers. Because the pages are not attached to the spine, the spine flexes far less, and doesn't act so much like the spring set into the hinges of a self-closing door, pushing the cover shut.

The RKM 200 allows this same trick to happen in a paperback book. The bad news is that all this machine does is bond the special tape to book covers and score the covers, and yet it costs $6,000: quite a bit to pay for a fancy tape dispenser. Planax being Planax, the brochures fail to make it clear that you also need to pay an additional $2,300 license fee for the privilege of using the patented system.

Covers that get the RKM 200 treatment should work in any binder, though they are intended for use with cold-glue systems (and using them with a hot-melt adhesive would likely defeat the purpose, as cold glues are more flexible). Of course, Planax wants you to use the the RKM'ed covers in the FII.

With the lay-flat adhesive and the lay-flat RepKover system working for you, those covers ought to lay *really* flat. As you will have paid $25,000 for the binder, $6,000 for the tape dispenser, and $2,300 for the license, you're certainly entitled to *expect* them to lay flat.

One other note to consider is that books bound with cold glue require drying time. The FII ships with a heated drying table, and the brochure claims that books will be ready for trimming after 15 minutes—though it couldn't hurt to give the glue a bit longer.

The cost-per-book of the special lay-flat tape is about 26¢. Five kilograms—about eleven pounds—of the book-binding adhesive goes for $135. This sounds like a lot, but if adhesive consumption is comparable to what my Bind-Fast does, that's about maybe 10¢ a bind.

Of course, I couldn't help but have the subversive thoughts that it just might be possible to dream up a way to glue a strip of thin strong cloth on either side of a cover spine that costs a lot less than six grand; and that you can apply cold glue with a $2.70 paint brush instead of a $25,000 binding machine.

But if what you are doing is cookbooks or music books or phone books or computer manuals that *must* lay flat, and you need to do relatively high-speed production runs, paying $34,000 for the hardware that lets you do it might actually make sense—and it seems clear this system does work. Do a quick search on the Internet for RepKover or Otabind, and you'll quickly find a multitude of articles and comments by people who swear by Otabind.

Obviously, I'm of two minds about the FII and RepKover. The system works, but it's *expensive*. Still, if you sell a thousand books for $40 each (now *that's* optimism) you'll pay for the hardware, cover your material-cost per book, and can start to make a profit.

Or you could shell out $34,000 and never sell anything at all. It's the *uncertainty* of the thing that makes it so exciting.

But if thirty-four grand isn't exciting enough for you, in the next section we'll discuss some gadgets for large-scale book-on-demand, where you can get the chance to spend some *real* money.

5. Specialty and Large-Scale On-Demand Printing Hardware

In the next chapter, we'll discuss getting someone else to do the work for you, and focus on contending with the various sort of book-on-demand printing businesses.

Even if you decide not to set up that basement print line and hire someone else to do your printing, even if you never print or bind any books for yourself, the material we have covered so far will stand you in good stead as you deal with outside shops. They simply use bigger, faster, more-automated versions of all the machines we've discussed, from printers to laminators to cutters to binding machines. However, there are a few intriguing pieces of hardware out there unlike anything we've discussed, and we might as well talk about them here.

On- Demand Sewn Bindings and Case Binding

Really high-quality hardcover (or case-bound) books are not glued together, but sewn together—Smythe-sewn, as it is called. A sewn book is more flexible and durable than a book that is merely glued together. One company has brought the advantages of sewn bindings to the world of digital, on-demand printing.

Kristec Automatic Book Sewing Machine

Manufacturer: Meccanotecnica
Distributor: Book Automation, Inc.
Approximate Price: $190,000

This machine takes in cut pages, which have been printed in exactly the right order, folds them, and assembles the folded pages into signatures. It then sews each signature and stacks the completed signatures together, and sews the gathered signatures into a completed book block. It does all this automatically and it's a lot of fun to watch it work. One version of the machine takes in the pages from a hopper which is loaded by an operator, while another is an "in-line" unit, meaning it takes the pages directly from a DocuTech or other digital printer. This machine is way beyond the dreams of a small-scale publisher, but it sure is cool. It produces very good quality sewn book-blocks.

Just incidentally, the printing order for pages intended to be sewn into signatures is entirely different than that for pages intended for an adhesive-binding perfect-bound book. In other words, there would have to be a different imposition for the hardcover and soft-cover versions of a given book.

On-Demand Case Binding Line

Three separate machines,
total price approximately $62,500
Available for lease for about $1,300 a month
Manufacturer: On-Demand Machinery

This system consisted of three machines that, with a fair amount of handwork, perform the following steps to produce a very sharp, professional-looking hardcover book.

(1) Machine one glues the printed cover to the front, back and spine pieces of bookboard that stiffens the finished case binding. (2) Machine two glues the cases to the previously assembled book block. The book block can be Smythe-sewn, adhesive bound, or held together pretty much any way. (3) Machine three compresses the spine and book block and forms the spine of the book. The whole process took about two minutes of labor per book.

I saw a demonstration of this system that included a Xerox DocuPrint 65 (a 65-page-per-minute printer), a perfect binder to form the printed pages into adhesive-bound book blocks, and a big guillotine cutter. There was even a specialized printer that did foil printing on the book covers. It was an entire line for printing and binding hardcover books, there in a trade-show booth. The gentleman at the booth estimated that the entire cost of all the booth equipment was about $200,000.

I should note there is another company, called the On Demand Machine Corporation, or ODMC, that has no relation to ODM, although ODMC is marketing machines that are similar in concept. They are discussed in the section below.

6. "Black Box" Printing Systems

The term "black box" refers to any gizmo that does its work by itself without intervention. The operator doesn't know or care what's inside the box. All the operator needs to know is that if you put the raw material in *this* end, the finished product comes out *that* end. For the purposes of this discussion, a black box is a machine that takes in paper stock, ink, toner, binding adhesive, reads a few computer files, and spits out a finished book without human intervention.

This is far from a new idea in book production. Most higher-end printers, binders, cutters, and so forth are designed to be used "in-line" with each other, with each machine taking up the output of the machine before it in line. The printer outputs pages directly into the binder, which drops the pages directly into the paper cutter. What's new is the idea of taking such "in-line" operation out of the printer's shop and putting in the back of the bookstore, or the university copy shop, or wherever. The other new wrinkle is to build such black-box bookmaking machines small enough and cheap enough to put into bookstores.

The *On Demand Machine Corporation* sells two machines, BookBuilder One and The Book Machine, that are all-in-one book printing units, intended for use in a bookstore or in a publisher's print shop. The web site (www.bookmachine.com) is rather vague. The only pricing information I could find was that the more expensive unit could be leased, starting at $1,600 a month, but that's old information that I couldn't confirm. OMDC also offers the *Book Finisher*, which takes pages and covers created elsewhere and binds them automatically. It seems to be a a more automated version of the binding machines we have already discussed. It does hot-glue binding, but also something called "sonic binding (patent pending)"

Another company, *Perfect Systems*, which is a limited partnership connected to *Marsh Technologies, Inc*, has some relation with ODBM. The situation is a bit complicated. The "Bookfinisher" ODBM offers is identical to a machine called "PerfectFinish" offered by Perfect Systems. Jeff Marsh, the owner of Marsh Technologies, is the vice president of ODBM. Perfect Systems does have operational prototype machines that make books, and there are videos of them available for viewing at their website. It would appear that Perfect Systems does have working machines—prototypes, but working machines.

Instabook, a machine sold by the Instabook Corporation, is another variation on the same "black box" idea. Feed the computer text file of a book in one end, and finished books come out the other. The machine, about the size of a office copier, includes a cover printer, a text page printer, and a binding system.

As with the units from ODMC described above, all I know about Instabook, I found out through the web site. To put it charitably, the web sites for these companies are not exactly overburdened with cold hard information. They haven't quite finished inventing the wheel yet. The websites are full of vague information and very old news reports, and the list of installed units isn't getting much longer. The revolution in in-store push-button book printing hasn't quite taken off as planned. It might be canceled due to lack of interest.

I haven't worked too hard digging up information about these machines. I don't know quite how close to available they are—or for that matter, how real they are. As best I can tell, all these companies are still at the prototype stage, with no actual machines yet available for sale.

Another reason I am hesitant about these all-in-one units is that they put all your production-equipment eggs in one basket. These machines are a Brand X page printer, a Brand Y cover printer, a Brand Z perfect binder and a Brand Q automated cutter stuck in a cabinet. If something goes wrong with the unit's page printer, that would bring the whole system down, because it's all interconnected. If you have separate machines for the printer, the cutter, the computer, and so on, you stand more chance of being able to do some work-arounds by borrowing a machine or getting part of the job done elsewhere if one machine breaks. (The machines offered by Perfect Systems are designed to be modular and allow hardware to be swapped in and out.) Furthermore, these machines all have various chutes and conveyor belts to transport the materials around inside between the various component machines. All that is just so much extra gadgetry that can go wrong.

It should be obvious by this stage of the game that I like gadgets and machines. But there are some situations where using a plain old human being is the best, simplest, and cheapest way to go. In this book, we have now discussed machines that print pages and covers, machines that bind books, and machines that trim paper. All these black-box units add to that combination is gadgetry to move the stacks of paper around, and sensors and software to tell the binder when to bind and the cutter when to cut.

That's the main thing that bothers me about these machines: they depend on a lot of complicated gadgetry to do very simple tasks, like picking up a stack of paper and putting it in a clamp. I can't help but reflect that it's the tendency of paper-handling equipment to jam easily that has made the all-powerful copier repair man into an icon of urban folklore, and the image of the hapless office worker vs. the recalcitrant copier into a cliché. (Bear in mind, you're likely dependent on the inventors of these machines to arrange for service.)

Thought of in this light, the black-box systems seem to me like nothing more than service calls waiting to happen. There the black-box bookmaker will sit at the back of the bookstore, under the slightly dusty banners reading NEW! IN-STORE BOOK PRINTING! And taped to the $65,000 black-box bookmaker is a hand-scrawled sign that has been there a few days, and was plainly written by someone both agitated and fed up. It reads OUT OF ORDER.

The bank loan used to pay for the bookmaker, however, is in perfect working order, and has to be fed the first of each month.

As best I can see it, these machines are there for the sake of clerks who are afraid to operate machines. The alternative would be to speak soothingly to the fearful clerk (who is on duty and has to be paid anyway), and then train him or her to run a laser printer, a binder, and a cutter, and assume he or she will know know to carry paper from one to the other. In short, these tasks are easy for people to do, and much harder for machines to do. You could attach the basic operating manuals to the sides of the machines just to be sure.

Still, turning every bookstore into a book publisher is a neat idea, and I'm being a bit hard on these machines, and the people who have plainly worked so hard on them. The videos of books being made available at the Perfect Systems website are impressive. I'd be delighted to be proved wrong in all this, and to find one of these machines cranking out deathless prose in the back of every bookshop. Just for now, however, I'd advise you to let someone else buy the semi-prototype black box bookstore printers. Wait just a bit until the machines are perfected and shipping.

In the meantime, let's find out how to get books printed on the machines the big guys have.

Chapter Nine
Using Large-Scale Book-On-Demand Services

There is a huge gap between the sort of small-scale, or table-top, book-on-demand printing that is the main subject of this book, and what might be called large-scale, or enterprise, book-on-demand printing, which uses hardware costing hundreds of thousands of dollars, or even several million dollars. However, as we shall see, it is very easy for the little guys to get books printed on the big guys' machines.

1. Access to Large-Scale Hardware

The printing of books in and of itself occupies a surprisingly small niche in the larger world of high-end on-demand printing. Oversized full-color posters, in-store advertising, fliers, junk mail, corporate newsletters, and all manner of other things are done via one form or another of large-enterprise on-demand printing. Books are close to a sideshow, with much (though far from all) of the book-on-demand work being done on hardware meant to do something else and adapted to the purpose, rather than on machines custom-built for book printing.

Lots of businesses sell small publishers printing and binding done on expensive book-on-demand hardware. Techniques, machines, and materials intended and developed for the bigger operators are available to and appropriate for smaller-scale operations.

2. Third-Party Book-On-Demand Business Categories

In general terms, "third-party" means that anyone other than you and your final customer. For our purposes, it means someone that you hire to do part or all of the job of printing and/or publishing.

A small-scale publisher operation can by copy a few computer files to a disk (or sending the files over the Internet) and send them to a *short-run book printer* which then puts its million-dollar operation to work for the book-on-demand publisher. For a quoted price, usually based on the page count, paper type, complexity of the job, and other variables, they will print however many books you order. (Longer "short runs" of a book, say, over a thousand copies, will likely be printed on a offset press, and not in a toner-based system.

You can also send those same files to a *Publishing Service Provider* (PSP) as discussed below. From a technical standpoint, this is going to be virtually identical to sending files to a printer. However, PSPs operate under a completely different business model than book printers. They are offering a service package, and the actual printing of books might be entirely incidental to that package.

There is yet another business model, which we'll call *digital warehouses*.They are set up for the express purpose of printing and shipping books for existing large and small presses.

All these businesses use super high-speed digital printers. Digital color printers crank out the covers. The books are bound and trimmed by the same processes as we have discussed already. The machines are bigger, faster, and more automated, but they work the same way.

Those machines do good work. The sample books I have from them are virtually indistinguishable from conventionally-printed trade paperbacks. It would be possible to tell the on-demand printed article from the "real" books, but only after careful examination, and the differences are largely matters of being able to distinguish one from the other, rather than a matter of the on-demand version being of lower quality. For example, some of the laminated on-demand covers had very slightly raised "ink" (toner, really), because toner sits *on* the paper rather than soaking in like ink. In similar vein, the hardback covers looked quite sharp, though some of the laminated covers on high-end book-on-demand hardcover books are too glossy for my tastes.

If you have a one-shot project, and you're more interested in getting books made, rather than in making books yourself, consider using one of these businesses to do the job for you. *You* would have to spend time learning how to bind books, but *they* already know how. *You'd* have to buy equipment. *They* already have it. If you do the job wrong, *you're* stuck with the result. If a professional print shop does it wrong, *they* will have to do it over for free.

3. Submitting to Third-Party Book-on-Demand Services

No matter how they do their business, all of these shops will want you to submit your work in more or less the same way. Generally speaking, they'll want you to send in one computer file for the book pages, and another for the cover. These days, most shops will want the final layouts of your book pages in an Acrobat PDF file, though they might accept—or prefer—some other format. Some shops want PostScript files. Some shops will take PageMaker, Quark, or InDesign files. Some of the PSPs will want the text in a word processor format, such as Microsoft Word or Word Pro or RTF, as they'll want to do the layout themselves. In any event, bear in mind that, unless you pay a lot of money to someone, no one else is going to read over your text. Every typo left in the text is going to appear in the book. Get everything right before you send it in. (Here again, I speak from bitter experience. And I'm willing to bet you've spotted a typo in this book, if you're reading an early version of the revised edition. Send your findings to typo@foxacre.com.)

For the most part, the files you submit will be the same files you'd use to print at home, though perhaps with minor modifications. Whatever the format, the business you send the files to will have very specific parameters regarding how they want the file prepared. There are a dozen possible variables that can be adjusted inside each of the "standard" file types. Follow the submission instructions completely and exactly. If you have questions, don't feel shy about asking. You are, after all, the paying customer. Besides, these instructions are usually in techno-gibberish, and it would be unrealistic to think the average citizen would know all about embedding fonts or switching off hyphenization or setting image compression.

One advantage of submitting a PDF file is that your pages are completely locked down. Except in very unusual circumstances, or unless you done a *very* bad job in creating the PDF, the final layout will look exactly like the pages you submit. This is slightly less true of a page-layout file. If your PageMaker file was set up with the print driver for printer X, and the print shop has printer Y, or if you did the Quark layout on a Macintosh and they use Quark for Windows, or if there is confusion regarding fonts, then the layout might "reflow" a bit, which could mean, for example, that a line or two of text could move from the bottom of one page to the top of the other. One possible result: reflowing can lead to the end of a chapter being truncated, with, say, the last two lines of a chapter vanishing altogether. You can get hit by different problems that will give you similar headaches if you submit a word processor file. Follow the instructions on formatting your files, and make sure you get a proof copy, and check the proof carefully. (Proofs are not an option. Don't try and save money by skipping them.)

The printer or service company will either want the files sent on disk, or uploaded via the Internet to their website. Most book files are unlikely to fit on a floppy disk. If you send on a disk, you'll almost certainly have to put the files on a CD-ROM or a ZIP disk, or other large-format disk. Just make sure they are set up to read the type of disk you send. If you elect to upload your files, be prepared for a long session, unless you have a high-speed internet connection. Book files can get to be huge.

At the end of the day, your files will get from you to them, and it will be up to them to produce your books.

Now it's time to examine the various types of businesses that produce books on demand in detail.

4. More On Publishing Options

Before we go on, a little truth-in-advertising. Versions of this next section have been through a lot of rewrites, and appeared a lot of places. What follows started out as a section of Chapter Two of this book. I excerpted portions of that chapter and rewrote and expanded it, and sold it as an article to Science Fiction Chronicle, a trade press magazine. I was asked to do an article on book-on-demand for the Science Fiction Writers of America's SFWA Handbook, and I based that article on the magazine piece, adding and updating once agian. Now I've taken that article, updated, revised, cut, pasted, and rejiggered it yet again—and come up with the following, which now goes here instead of in Chapter Two. So if some of this seems a bit familiar, maybe it's not you. Maybe it's me.

***

To overstate the case a bit, printing books, is easy—or at it's a lot easier than publishing books. Write your words, proofread very carefully, lay out your text, and design your covers. Then send disks with the pages and cover designs to the printer, add money, and you're done. Alternately, watch the ink-let and laser printers in your basement do their work, spend a few hours cutting and gluing and laminating and trimming, and *then* you're done.

However, there are likely readers of this books who have concluded that printing their own books is more than they are ready to take on. And as we saw in Chapter Two, *publishing* the books involves a lot more than just printing.

There are businesses ready, willing, and at least claiming to be able to take on any or all of the tasks of publishing. These businesses range from real publishers large, small, and microscopic, to book printers and digital warehouses, to businesses best described as Publishing Service Providers, and then on over to neo-vanity presses (many of which retain many of the sleazy habits of the old-line vanity presses), and finally on over to out-and-out scam artists. We'll discuss them all in detail in the remainder of this chapter.

5. Conventional Publishers

Of course, the best way to get books printed and/or sold to readers is to get someone else to do the work, and make him pay you for the privilege. In other words, consider submitting your work to a real publisher—maybe even a small press that uses book-on-demand printing. (While it is of little importance to you, the writer, how these new publishers print their books, many would not exist were it not for the new printing technologies that make it cost-effective to print only a few books at a time.) What we're talking about here is a publisher that performs all the functions discussed in Chapter Two, and pays you for privilege of printing your work, and pays you again when more copies get sold. It's nice work if you can get it.

Before you do submit your work, check the publisher's Writer's Guidelines. The publisher might call this information their Submission Rules, or Author's Information, or something else, but, by whatever name, the publisher will have such rules. In sheer self-defense, all publishers come up with guidelines, telling you what sort of books they publish, what they *don't* publish, precisely how to submit, and, more often than not, adding a plea that you not ignore the guidelines. These guidelines will be found in various reference works at any decent library, or on the publisher's website, or will be available by mail upon request. FoxAcre Press has its guidelines right on the website. *Always* follow the publisher's guidelines. If you're trying to make the publisher do nice things for you like publish your work, don't start off by ignoring instructions.

6. Short-Run Printers and Digital Printers

These are certainly the easiest businesses to understand. You shop around for a printer with expertise in the sort of printing you want. You tell them what you want printed. They quote you a price. You get prices from three or four companies, and do what you can to evaluate the quality of their work (maybe by getting sample books). You send them the files they need, and they send you back finished books. Some shops are set up to do fifty copies, some to do fifty thousand, and others do every number in between. Prices vary widely.

Recently, just to get an idea of prices, I requested estimates for printing copies of my book, *Orphan of Creation*. I queried both on-demand printers and conventional offset book printers. (I should note that, at 340 printed pages, *Orphan* is going to be

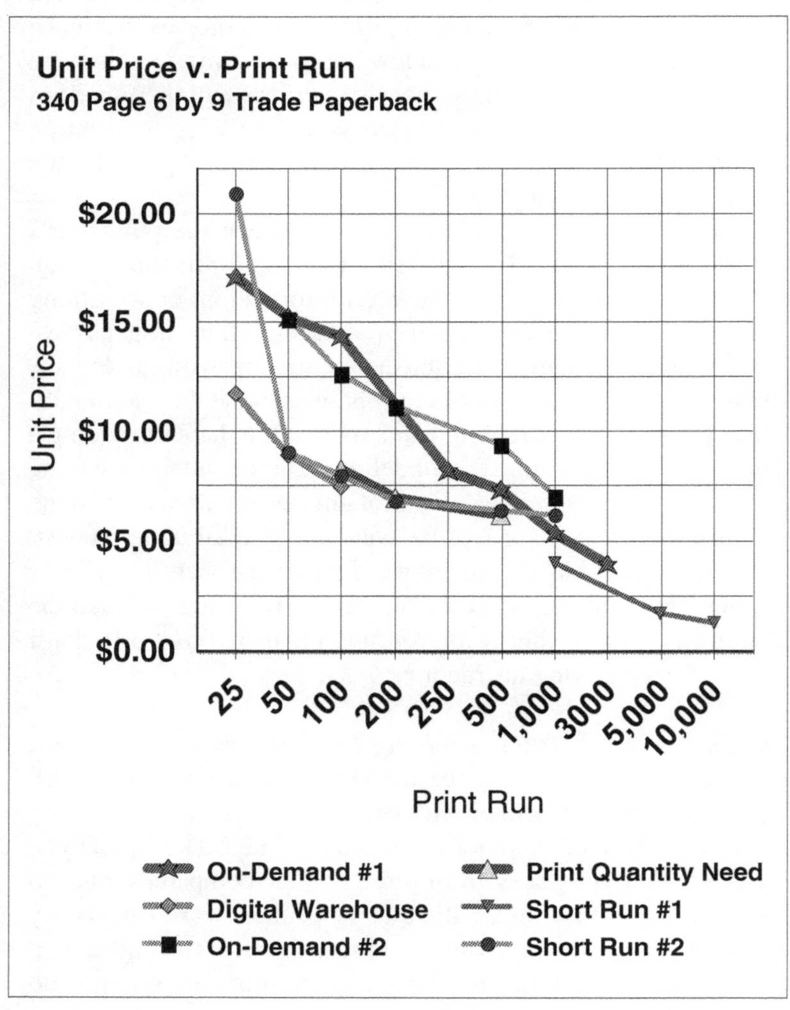

well toward the high end of what's practical in terms of price for book-on-demand. Book-on-demand printing is charged for by the page. The more pages, the more it costs.)

The unit cost of a book printed 25 at a time could easily be fifteen times higher than the unit cost of the same book printed 10,000 at a time. I received quotes as high as over $20.00 per unit for 25 copies, down to $1.34 a copy for 10,000 copies. Prices also vary widely at a given print-run size. As show in the accompanying graph, the cost per book at the hundred-copy print run ranged from $7.96 to $14.35 per copy. (Moral: Shop carefully.)

Also note the figures in the chart reflect all set-up costs. If they charge you $100 to set up the book, and $400 to print 100, your unit cost is $5, not $4.

As prices change over time, I have not identified the companies I queried, but instead labeled them by business type. Also note that hardcover books are available from nearly all of these companies—for much higher prices.

There may or may not be a relation between price and quality. Someone might quote you a sky-high price because they don't do that sort of work and they don't know how to price it. In such cases, you'd be more likely to correlate bad quality with high prices. Or else a vendor might simply be trying to chase off a job they don't want by quoting an unrealistic price. Too low a price might be a sign of desperation, of a business that will take any job at any price because they need the work. Do some homework. Know more about the printer than the price they charge.

Para Publishing offers reports on how to buy book printing through their website. Also see www.bookmarketing.com for a list of printers.

7. Publishing Service Providers

The business category I call Publishing Service Providers (PSPs) stands somewhere between vanity press and conventional publishers, with a dash of print-shop thrown into the mix. In a typical deal, for about $100 to $400 (and sometimes the fee is waived), many publishing service providers will store a copy of your book on their computerized printing system, do order-taking and fulfillment, print and ship books as orders come in, and then split the proceeds of any sales with you.

Typically, these shops charge you for layout and printing, maybe give the book an International Standard Book Number (ISBN) (thus getting it into *Books in Print*), offer it in their catalogs, and provide a way for an author to sell a book without starting a publishing operation. They might or might not get the book placed with any of the on-line booksellers. The entry-level fee will get you a very basic book with a generic cover. Most shops charge more for various "extra" services that you'll probably want, like a custom-designed cover. Generally speaking, they do no marketing or promotion—and the honest ones don't pretend to do it.

This approach has the advantage of getting the sales and fulfillment job off your back, and also keeps the stacks of books out of your basement. Those who want the book can order from the web address or toll-free number of the publishing service company, or maybe even from online bookstores. You, the writer, thus let the publishing service company worry about shipping and storage and credit card numbers and so on, but it's up to you to promote your work and get people to order the book.

These businesses are legit operations. They are doing something that looks like vanity press, but it isn't really. The service they offer can make sense for those who don't want to deal with the trouble of printing and binding their own books, but do want to have their work available for sale from a commercial source.

It might also make sense to use this sort of service for printing specialized textbooks. A teacher could write the name of the book on the board along with the PSP's toll-free number and tell the students to order the book from the PSP. These shops could also be the logical home for specialized texts and reference books on extremely focused topics, for good fiction that is aimed at a small and specific audience, and so on.

As good as all that sounds, such pay-to-publish services violate that cardinal rule of publishing I banged away at so hard in

Chapter Two: *the money should move toward the writer.* That still bothers me.

8. Neo-Vanity Presses and Scammers

At least some PSPs act more like high-tech vanity presses than publishing service providers. Call them neo-vanity presses. The neo-vanities companies make the lion's share of their money from fees paid to them by the writers, rather than from sales of books to readers. They make most of their living by receiving fees from writers who fail, rather than by sharing money with writers who succeed. One such firm claimed to earn half its money from fees paid by authors, and half from the actual sale of books. That sounds like a suspiciously *high* percentage from book sales. (Or maybe it means people are wising up, and they aren't receiving that many fees from writers these days.)

These shops are where the amateurs go. As the old-line vanity presses decline (and I'm glad to see them in trouble) the neo-vanity presses are more and more where all the bad thrillers by dentists wind up. They are the new home for incoherent political rants, cheesy ego-driven memoirs, fervent religious tracts, and bad poetry. Assuming your check clears, there are no editorial standards or quality controls. Readers will no doubt quickly learn to be leery of books published by these companies. Result: if your book is "published" by a neo-vanity press, so far as the reading public is concerned, it will be lumped in with *The Purloined Bicuspid* and *Odes to Lost Pets*.

A further downside is that these publishers, generally speaking, require you to publish under their imprint, and not under your company's name. (There are exceptions to this—if you pay a higher fee.) But if you sign up with All-Our-Books-Are-By-Amateurs.com Press, and that's the name that will appear on the spine of the book. None of the real names of PSP "publishers" are quite that much of a giveaway, of course, but books produced by PSPs usually are pretty easy to spot.

Be aware that some of these services don't even assign ISBNs to their books, which means no bookstore or online bookseller will carry them. The *only* places it is possible to order books in such cases is through the author (if he or she has purchased a stock of books) or directly from the publishing service.

You could choose to go with such a PSP and be entirely happy with what you got, because you knew exactly what were getting yourself into. But there are plenty of wannabes out there, and plenty of out-and-out scammers who will take their money and run. Rather than charging for a service and then delivering it, they'll out-and-out lie, take your money, and then not deliver, or else cut so many corners on what they do deliver that they might as well be stealing. We are, of course, talking neo-vanity presses. These shops did exactly what the PSPs do, but they lie and say they can do more. Remember the old rule. *If it sounds to good to be true, it probably is.*

A book produced via book-on-demand isn't likely to sell tens of thousands of copies, or get you on the best-seller lists, or make you rich. Any PSP that suggests otherwise is not likely to be playing straight on other matters.

It would probably be safest, for example, to assume that you cannot believe a word any of these shops say about promoting or marketing your work. Either they won't really do promotion at all, or else they'll promote in a completely useless way, going through the motions just enough to avoid a lawsuit. Even if they wanted to promote books effectively, they very likely don't have the time or resources or knowledge to do it. The short form: you will be on your own when it comes to marketing. See Chapter Ten for a brief discussion of this subject.

As noted above, the honest shops won't pretend to promote your work. In large part, that is the difference between a PSP and a neo-vanity. They both do more or less the same physical tasks: printing and shipping books. It's just that a neo-vanity press pretends that it is doing more. If they pretend too hard, over-promise too much, and under-deliver thoroughly enough, what you've got hold of is not a business, but a scam.

The borders between legit PSPs, neo-vanity presses, and scam artists are badly blurred. All use the same technology, all offer seemingly similar services—and one company might well be playing all three roles. It's up to you to shop carefully. Don't go with the first service you find. There are lots of deals out there. Find the one that fits your circumstances best.

Remember that it can be in the company's interests to blur the lines as to what role they play. Many PSPs, for example, try to look like publishers, and their ads urge you to "Let Us Publish Your Book." How can you tell a publisher from a PSP and either from a scam shop?

Here's a few rules of thumb—a publisher advertises its books for sale to the reading public. A printer advertises its printing services to publishers. A PSP advertises its accurately described services to the writer. A neo-vanity will also advertise to the writer, but will sell too hard, and likely make a lot of vague statements that sound like they mean more than they do. A scam shop will offer deals that are clearly too good to be true, make implausible promises, and claim to achieve impossibly good results. Head to the company website, and see what they are selling to whom and how plausible their claims are.

There are a large number of businesses set up as on-demand book printers, or as publishing service providers, or both. There are so many of these businesses, and they come and go fast enough, and change their service plans so often, that there is no point in including a table comparing rates and policies. Visit *www.bookmarket.com/ondemand.html* for a list of such companies.

9. Contracts

Whatever sort of service you choose, there will at some point be a need for a detailed agreement, setting down in writing who is to do what and how much must be paid and when. Perhaps the most important advice I can give in this chapter is this—read the contract carefully.

Most printing contracts are going to be very straightforward agreements, all about delivery date and printing specs. But publisher's contracts are inherently more complicated, and PSPs and neo-vanities are enough like conventional publishers that they offer contracts that are variants of standard publishing agreements.

Authors have gotten into the habit of assuming the don't have much leverage with publishers—an accurate perception for the most part. But if you are *spending* money, instead of hoping to be paid money, you are the customer. You can take your money and go across the street. Don't accept a take-it-or-leave-it contract. If you don't like a particular contract clause, but the PSP says their contract is non-negotiable, fine—don't negotiate. Walk away instead. Remember, there are plenty of competitors.

Don't deal with a company that doesn't offer some sort of written detailed agreement. Don't accept a contract you don't understand, and *do* be determined to retain control over your own book's rights. There are insanely bad contracts out there, which in essence ask you to pay for the right to surrender all rights in all forms for all time.

Insist on a clearly limited grant of rights. Basically this means a contract wherein you grant rights for a specific period of time, for specific types of editions, and maybe for a specific geographic area. For example, "the author herein grants for a period of three years the right to produce a trade paperback edition of the Work in the North American market" wouldn't be a bad grant of rights. "The author herein grants for the full life of the copyright all rights to the Work in the World and throughout the Universe, to produce editions in any and all formats, including formats not yet invented at the time of this agreement" would be a very bad grant of rights. Under current copyright law, you wouldn't get the rights back until you had been dead for 75 years. (And I'm not making that grant of rights up, either. A friend of mine was offered a contract with almost exactly those terms. He said no.)

Make sure the contract requires the PSP to provide regular statements of account, and regular payments of amounts due.

One contract problem has, ironically enough, been created by book-on-demand technology itself. On-demand printing makes it absurdly easy to keep books in print. Most publishing contracts have a reversion clause, clearly defining the circumstances under which the rights granted to the publisher are returned, or reverted, to the author. Reversion need not take place at a specific time. The contract might instead state that the right will revert when sales drop below a certain rate—or when the book is no longer available for sale. In the old days, that was a reasonable proposition, because the publisher would have to print a lot of copies and keep them in the warehouse, and available for sale in order to keep a book in print. In the age of book-on-demand printing, this reversion clause is far from reasonable.

A reversion clause that merely requires a book be "in print" or "available for sale" means that a publisher could use on-demand printing to keep a title in print, and thus retain the rights, forever. The publisher might not even need to retain a physical copy: all that would be needed was a copy stored on disk somewhere, ready to print.

The ideal situation from the standpoint of the author would be the ability to get the rights back immediately upon demand at any time, though it's unlikely any publisher will be quite that generous. The publisher will want to have complete control for a clearly defined period that is long enough to give them a fair chance to make some money, and preferably longer. The time-limited grant of rights is a compromise between these two positions.

Note that the "good" grant of rights clause described above does not necessarily mean the rights will revert in three years. Typically, at the end of the initial rights period, the rights renew automatically until and unless either party cancels. If you're happy with things as they are, you take no action and the PSP retains the right to publish. If you're unhappy, you write and tell them to revert the rights. The details may vary, but with a time-limited grant of rights, you should be able to renew or cancel with reasonable notice at any time once the initial grant-of-rights period ends.

In the past few years, more than a few PSPs have promised that their books would stay in print "forever"—and then gone out of business. In fact, several came into being and then vanished in the time it took to research and write the first commercial version of this book. Check to see what any contract says about what happens to rights you have granted if their business dies. It might be that they can sell the right to publish your book to their creditors, or to whoever buys their company. That's not necessarily a bad thing for you, but in an industry this volatile, it's important to take into account that the business you are signing with might fold or be merged into some other business. Know what will happen in such circumstances.

10. Digital Warehouses

Outfits such as *Lightning Source Inc.* (owned by Ingram Distributors), *Replica Books* (owned by Baker & Taylor) and *Digitz.net* (formerly independent, now part of BookSurge), are rather different operations from PSPs as described above. And, here again, things are evolving fast. New services could start up, or old one merge (again), very quickly.

Such businesses are a cross between a digital printing service and a warehouse. Many big commercial publishers use these services to get early rush copies of titles made up, or to keep older titles in print. Small presses use them a lot. More and more trade paperback reprint are being printed by one of them.

The following discussion focuses on Lightning Source, but the basic business model should be the same for any of these services, with some variance in fee structures and shipping policies and so on. Double-check the rates and fees described. Things change.

Lightning Source (LSI) in essence acts as an on-demand digital distribution center (call it a digital warehouse) for publishers. Publishers send in books as computer files. LSI stores the books in electronic form. They print and ship copies whenever orders come in, using Ingram's normal ordering system. LSI titles are also listed in Baker & Taylor's ordering systems. So far as the bookstore doing the ordering is concerned, it doesn't matter that the warehouse contains, not a stack of pre-existing books, but instead a merely text on disk that is printed only after the order arrives. The orders go in, and the books come out: a virtual warehouse. Titles in the LSI catalog appear automatically on

Amazon.com, BarnesandNoble.com, BooksAMillion.com, etc.

With a book stored in a such a system, all a publisher has to be is someone with a few International Standard Book Numbers. There is no need to do home book-on-demand printing, PSP printing, or conventional printing. In theory, a publisher does not need to print anything, but instead merely collects a fee whenever someone places an order at a bookstore or phones or mails in an order, or clicks on the order button on the web page, causing his or her book to be printed out and shipped. Theory sometimes conforms to reality: I can report the real sense of satisfaction I get each month when I receive a check for doing no work at all.

The downside is that I get much less money per book when Lightning Source sells the book for me. If I sold a copy of this book to you directly at the cover price of $21.00, I'd get to keep the whole $21.00, less the cost to me of printing the book, roughly $3.35, leaving me with about $16.65 to pay the electric bill and the gas bill and the other bills, and compensate me for the work of writing, illustrating, printing and binding the book. If I sell through Amazon.com, Lightning Source handles the transaction for me. They sell the book to Amazon at a 55% discount, meaning they sell the book for 45% of the cover price, which comes to about $9.45. LSI then deducts their printing cost of about $4.05 from that amount. This works out to my receiving about $5.40 per copy sold this way. Thus, I earn about $11.25 less per copy than for books I sell direct. I also get the money about three months after the sale, as opposed to receiving at once, as I would with a direct sale. But still, it's money for no work, and it's money for sales I would not make at all without LSI.

LSI also provides straight short-run printing services for publishers. This service can be used to print Advance Reading Copies or finished books. LSI charges a set-up fee for each title, based on the number of pages in the book. Typically, this fee will run about $125 to $175. LSI also charges by the page (plus a fixed charge per unit) for printing books. This works out to about $4.00 to $6.00 per unit for the average novel. While LSI will print one or two copies at a time for bookstore orders, for non-distribution titles they have a minimum initial short run printing order of 100 units, with a minimum of 25 units on subsequent orders. They have a minimum first order of 25 copies for short-run titles they distribute. Thus, a typical charge for 100 copies of a given title might be $150 for one-time setup plus $500 for the books themselves, or $650.00, or $6.50 apiece.

11. Specialty On-Demand Book Printing: Hardcover Books and Picture Books

This book has focused almost entirely on doing trade paperback editions with black text and perhaps some sort of simple black-and-white illustrations between two covers—in other words, exactly the sort of book you are holding in your hands right now. There are, of course, other sorts of books. I have been asked many times if on-demand printing could be used for hardcovers, and/or for full-color books, such as children's picture books. The answer these days is "yes"—so long as you're willing to pay, and so long as you let someone else do the job.

As discussed elsewhere in the text, it is possible to create hardcover books at home. I have chosen not to spend too much time on the subject in this guide because the goal here is to produce multiple copies of good-quality books cheaply, easily, and quickly. While there are relatively straightforward ways to do hardcovers, even the simplest case-binding technique requires so much handwork and time that it rapidly becomes impractical to produce such books in any number. The additional time and materials also send the price you'd have to charge for such books through the roof. Besides, there are any number of books on how to make hand-made hardcover books. *Dover Publications* stocks many titles on this subject.

Even if you're not planning to sell your books, you should think carefully about how much the books will cost you. Even if you're giving copies of your book away, working up a rough idea of how much you'd have to charge for the book if you *were* selling them is a useful measure how much effort you're putting in.

Many of the big shops discussed above offer books printed

on-demand with hardcover bindings, but the costs are much higher, and, because of discounts, shipping costs, and so on, higher costs to the publisher have a *big* multiplier effect on the final price. Let me offer one example. I looked into what it would cost to do a hardcover edition of a book that I sell in a trade paper edition for $13.50. The book is about 200 pages. My unit cost for printing the paperback version is $3.69 or so. The hard-cover cost per unit would be $7.94. According to the formula I use for costing and pricing, and figuring in the higher set-up costs, in order to more or less break even, I would have had to set the list price at a bare minimum of $31.00 for the hardcover! Even that cover price is a bit on the low side to cover costs. I might have had to go as high as $37.00. That price might be practical in some markets, but not in the ones *I* sell to.

Obviously, the unit cost drops if you do a longer print run, but I suspect you'd have to get up to a print run of at least a thousand before the unit cost of a hardcover would become at all reasonable.

12. Full-color Books on Demand

The situation on doing full-color books yourself is different. Things still get very pricey—they just get pricey in a different way. For starters, given the current crop of color ink-jet and laser printer available, I don't see any practical way to do production runs of color picture books at home. As we have seen, the images produced by these printers aren't robust enough to stand up to much abuse, and the typical picture book takes a *lot* of abuse. Then there is the issue of doing double-sided printing with color. Unless you were using dollar-a-sheet photo-grade paper, you'd unquestionably get unacceptable amounts of bleed-through. The paper would simply become saturated with ink, and you'd see the picture on the back of the page you were looking at.

You *might* be able to get away with doing text-only on one side of the sheet, and images on the other, photo-grade side of the sheet. Thus, each pair of facing pages would have text on one side, and the picture on the other. But you'd wind up with the two sides of the page having distinctly different textures, which would make the book seem odd to most readers. You could eliminate that problem, and protect the pages better, by doing a double-sided laminate of each page, but that would make the pages much too thick, make the inside pages *too* glossy, possibly introduce binding problems, and probably produce a book that was difficult to open and handle. Plus, given the high costs of ink-jet ink and photo-grade paper, plus lamination, the materials costs on a 32-page (sixteen sheets, printing one page per side) 8 inch by 10 inch book would probably run somewhere around $10 to $16 a unit!

Using a color laser might eliminate some of the problems, and (perhaps) cut the materials cost, but it would introduce new problems as well. Many color lasers produce a somewhat odd effect, in which some colors get printed with a very glossy surface, while others have a matte appearance, and areas without color printing are less reflective still. The effect is not noticeable when the page is viewed face-on, but if the page is held at any sort of angle, it is very pronounced—and of course the printing would be rather fragile on most paper stocks. Of course, you could solve the problem by doing double-sided lamination with a laminator and film designed to contend with fuser oils, but then we're back to binding problems and pages that are too thick...

In short, as things stand right now, I don't see any practical way for anyone to print full-color books at home.

The alternate solution: shop around with the Publishing Service Providers—but reach for your checkbook first.

(Note: as this book goes to press, this sort of service is just starting to become available, so look for changes in services offered and prices.)

Xlibris has announced a program for doing true print-on-demand picture books. As of this writing, the set-up fees start at $999, and keep right on going up to $2,499, depending on how complex the book is. Xlibris produces the books as they are ordered, shipping within ten business days of the order. Based on what I could see on their webpage concerning author reimbursement, you'd have to sell at least a few hundred books to break even. It would appear that you are locked into their pricing schedule, and for quanities under 100, the minimum cost of books sold to the author is about nine dollars. Given the possible difficulties of color laser printing described above, I would strongly suggest getting a sample book before proceeding.

RJ Communications, through their booksjustbooks.com website, offers an alternate approach. They use a more conventional printing process, require eight to 10 weeks for delivery, and a minimum order of 1,000 copies. However, the unit cost can be as low as $5.50, and you can set your own cover price—Booksjustbooks is acting as a printer, not a publishing service. The quality of the sample book I saw was excellent, with no sign of the variable gloss produced by some color lasers.

Xlibris acts as publisher, and takes a cut of every sale, while booksjustbooks is acting as a printer, not a publisher. They drop the books at your front door, and that's that.

As of this writing, Imprintbooks, a division of *Booksurge*, reported that they would be offering full-color books through what they called CMYK Books "soon," with a set-up fee of $499. In some ways, the services offered seemed more limited that those of Xlibris, and more flexible in others.

Bear in mind Xlibris and Imprintbooks are selling "publishing" services. These may or may not be right for you. There are lots of shops out there—large and small, local and national, that offer printing services. If you are simply looking for someone to print full-color books for you, you will almost certainly be able to come up with a better deal than what you'd get from the PSPs.

At the 2002 Book Expo America, Xerox demonstrated a very impressive system for doing full-color books, with both the color pages and the color cover being printed by the same Xerox printer, the Xerox 2060. The sample book they distributed looked very good, with only minor traces of the variable gloss problem described above. Prices for the full system, including the binding machine and paper cutter, "start" at $200,000, according to one press release. The printer and its front-end control are by far the largest chunk of that two hundred grand. That might be a bit pricey for some readers of this guide, and even some print shops.

There are other color printing systems out there, and more on the way. Therefore, while I have found only limited availability of full-color book-on-demand services, look to see them expand in the future.

To sum up: when it comes to third-party book-on-demand, you can get just about whatever you want, for a price. And you'll get what you pay for—if you shop carefully.

Chapter Ten
Prototyping, Production and Business Decisions

In this quick final chapter, we'll walk through a few of the blindingly obvious common-sense procedures I rarely follow myself, but should, and then talk, albeit briefly, about the business side of book-printing and publishing.

Those not interested in selling their books can skip the end of this chapter. However, even if you're planning to sell on a non-profit basis—for a church, or a school, for example, it's in your best interests to handle that non-profit on a very business-like basis.

1. Final Checks and Prototyping

Once you have your page layout complete, and have your cover more or less under control, it's time to print, bind, and trim a few test copies, or prototypes, of the book. Make your surprising discoveries about your book before you go into full-scale production.

Confirm that your scoring marks are accurate, that your book isn't too thick to fit in your binder, that the binder isn't going to melt the lettering off your cover, that the inside margins are wide enough to allow for the final trim, that the cover is long enough to wrap around to your planned margins.

Most glues, including some thermal ones, get stronger if they are allowed to cure and set for a while. Give all your inks and varnishes and glue time to set, and then sit down and look over the book, and *treat* it like a book. Flip through the pages. Make sure the spine holds together. Note the four or five typos that will jump out at you, and go back and fix them. Put the book on a shelf and step back a pace or two to see if the spine lettering is legible. See if the front and back cover are properly aligned with the spine and the outer edges. Compare two or three copies of the book and note how they are and are not consistent, and whether it matters or not. Set the book to one side, and look it over the next day, after a good night's sleep, to avoid being deluded, either by euphoria or gloom. Make the fixes you need to make, and *then* print some books. Don't do it the other way around.

2. Production Printing: Keeping Organized

Do detailed notes on your procedures for printing the book, with an eye toward the time—a week hence, a year hence—when you'll need more copies of the same book. Keep a file on each title you print, logging dates and sizes of print runs, what typos you have fixed, changes in print procedure, and so on. Try and set up a file that someone else will be able to work from, if for no other reason than to be sure you can understand your own notes later on.

Avoid a crisis three years from now when you suddenly need five hundred copies of the book and you can't remember the name of the font you used, or where the final version of the layout is on your hard drive, or which cover stock or page stock you need. And, of course, keep back-up copies of all your work.

Find a good way to keep records of corrected typos and updates so you know what you've done and what needs fixing. Choose a policy and stick to it as to how you manage updates of your various files—the manuscript from your word processor, page layout, and even your PostScript and PDF output files are likely to lurch out of sync with each other. If you fix the PageMaker file without correcting the WordPerfect file you used to do the PageMaker layout, or the PostScript or Acrobat file you generated from the PageMaker file. you could end up with a hopeless muddle if the PageMaker files is destroyed. Usually there is a way to export a text file back out from the pay-layout program. You might want to use that feature to avoid having to do all your updates in multiple places.

Keep backup copies of the word processing files and DTP files and graphics files for your inside pages, and for your cover.

There are now various brands and formats of removable drives that can hold a lot of data on the cheap. ZIP drives use $10 disks that hold a hundred megabytes of data each.

CD-ROM drives that let you "burn" your own CD-ROM disks are available for less than $200. Being able to store your book's text and layout information on a disk that ought to last a hundred years is probably worth more than that. The disks themselves are now so cheap that they are nearly free. Don't risk the loss of a year's—or even a day's—work because you don't want to use up a 25¢ blank CD.

It might be smart to put backups of all the files for each book on one disk, one disk per title. You could easily keep complete sets of files for each draft of even a very long book on one disk.

Above and beyond copying onto a disk of whatever sort on a regular, routine basis, take the time to make a complete set of all of your work and put it someplace outside your house, like a friend's house or a safe deposit box. If your house burns down, you shouldn't find yourself losing valuable business assets that could have been preserved on that 25¢ disk.

One point to think on is what your normal press run should be. Printing literally on demand is probably not the most practical way to go, simply because it would be a nuisance to trot down to the basement to print one copy every time an order comes in. Nor do you need a hundred copies of each title on hand at all times.

In most cases, it would make sense either to print up a set number at a time, or periodically to print up enough copies to maintain stock levels. One point of book-on-demand is to keep inventory down. Inventory ties up money, takes up room, and can increase your tax bill. (It might make sense to let your stock run out just before tax time each year.)

Find a print-run that uses your time efficiently, that keeps enough books in stock, and yet doesn't bury you in inventory.

3. Taking Care of Business

One of the biggest changes in book-on-demand printing since the last major update editions of this guide was the expansion, and then partial (in some cases near-total) collapse of various businesses that used book-on-demand printing technology. Most of these collapsed businesses can be filed under the category "started up by well-meaning people who didn't know what they were doing," with a few under "unbelievable stupid."

I have had occasion to look at a business plan or two, and/or to correspond with a few people who read previous editions of this book and set out to start their own publishing business. What scared me most about these folks was how much they didn't know they didn't know. I've written books for a living since the mid-1980s, and I've been running a publishing business for about three years now, and I *still* don't know exactly what I'm doing. (I like to think I've got nearly everything nailed down except the part about selling books.) But at least I am aware of my ignorance.

Anyone who has read this far in this edition of the book will have noticed the number of businesses reported as dead, or missing in action, since prior editions of the book. Many of these are Ghosts of the Internet—websites that are still live, touting businesses that are plainly dead. The website headlines quote four-year-old news stories, or urge you to "Visit Our Booth" at a trade show that happened literally in the last millennium. They list "new titles" with copyright dates of 1999. It can get a little creepy.

So please. If you're considering the idea of starting a business using any version of book-on-demand technology, do your homework. Study how the business you're interested in works. Identify competitors. If no one else is doing what you want to do, ask why. Maybe you're the first to get the idea—or maybe you're the last to find out what's wrong with it.

This is not to say you should not start a business based on book-on-demand printing. I did. (Someday it might even break even.) All I'm saying is know the risks, know what you can afford to do, and know all you can about the ways things are done in printing and publishing before you jump in with both feet.

There are literally thousands of books and computer programs and seminars and magazines and so on concerning the subject of running a small business. Make use of them.

4. Number Crunching

While we have talked business here and there in the forgoing chapters, the intent of *this* book is to provide the hard-to-find or not immediately obvious or unavailable information specific to on book-on-demand printing. So, now that we know a bit more about putting books together let's look another look at business tasks specific to publishing, stuff that won't appear in the standard small-business texts.

One very obvious suggestion: keep records of how many books you sell. Less obviously, track where, when, how, and to whom you sell them. If you find that certain books sell on the weekends, or toward the end of the month, or before a particular holiday, or not at all in February, you can use that information in future to fine-tune your print runs, focus your marketing efforts, and save wasted effort. The information can also help you get to know your audience, and have a better feel for what will and won't work in future.

But there is another sort of keeping count that can be just as important: the costs of doing business. It does you no good at all to sell thousands of books at $44.32 each, if it cost you $44.33 per copy to pay for raw materials, utilities, shipping, labor, and so on. Track your costs. Base your business decisions on the numbers, not on what you hoped or thought or wished the numbers would be. (Twenty or so years ago, I helped drive one family business into the ground by not following that rule.) Bear in mind that every business starts out losing money, so be optimistic. On the other hand, quite a few business never *stop* losing money, so be careful. As I believe W.C. Fields once put it: "If at first you don't succeed, try, try again. Then give up. No sense being a damn fool about it."

5. Fulfillment and Fulfillment Services

"Fulfillment" is a term used, not only by touchy-feelie self-help writers, but by people who take and ship orders. To pack and ship a completed order out the door and to handle all the financial details surrounding it is to "fulfill" that order.

Have a nice, clear, standardized way to manage your orders. Whatever way orders come in, via mail, email, phone, fax, or face-to-face, make sure they are managed in an orderly and standardized way. You might want to work up a checklist, and make sure you work against it with every order.

There are fulfillment service companies that will do all or part of your fulfillment work. This can be done several ways. You might ship your stock of books to their warehouse, refer customers to them, and let them handle it. You might retain your stock, and rely on them to call and tell you when an order comes in, so you can ship the books. There are companies that will staff your toll-free number, or manage your mail orders, and so on.

For 99 percent of the readers of this guide, this is far more service than you'll ever need—and the companies in question charge, sometimes quite handsomely, for their services. For the one percent who want to go this route, visit parapublishing.com, and search on "fulfillment." For everyone else, I'd strongly suggest you do your fulfillment yourself. It's pretty easy—if you've got the tools.

6. Fulfillment Software

For someone sending out a few copies of a self-published book on a semi-casual basis, keeping a paper list, and maybe a list of addresses on three-by-five cards, might be all tha t is required for fulfillment.

Moving up one step from simple paper record keeping, there are also plenty of general-purpose accounting programs out there, and most of these will do the trick for tracking expenses, keeping inventory, and so on. These programs are so cheap, and so good, that it only makes sense to spend the extra fifty or a hundred bucks after the two thousand you've spent on the computer.

For a small book-on-demand operation that's only managing a few self-published titles, such a general purpose program will do fine. If you are the only author you publish, you don't need to worry much about royalties.

However, a book publisher who puts out multiple books from multiple authors and sells to a large and varied group of customers (say, for example, some at full price via direct mail and some at 50% discount to bookstores) is in a much more complicated situation. Such a publisher has to keep track of inventory, track the royalties paid and the royalties due to authors, handle backorders, deal with differing discounts to differing accounts, track unpaid invoices, maintain a mailing list, generate mailing labels and invoices, and so on. A standard off-the-shelf small-business bookkeeping package is unlikely to handle these sorts of publishing-industry-specific tasks very well.

A good publisher's software package should take care of all of the above functions, and more, in the background. Every time an order is entered, the program should calculate the royalties for the sale, subtract the order from inventory,print a shipping label, and note down in the mailing list database that customer X purchased a book by author Y on date Z. These programs have other bells and whistles, but you get the idea.

I have located five specialized programs that perform these functions. There are other publisher's fulfillment programs out there, but they are for vastly larger operations. Most of the programs listed below are overkill and then some for a book-on-demand publisher. Because these are complex and expensive programs that not every book-on-demand publisher will need, I will limit myself to a brief discussion of basic product information rather than exploring them in depth. These are leads, not reviews. I should note that I had to add or update material on four out of the five products listed between two versions of this book. Things change fast, so do your homework and make sure you have the most current information.

Because they are not sold to the general public, but to a speciality market, most of these programs are wildly expensive, compared to prices for mass-market software. I list the happy exceptions to this rule first.

Adams-Blake Publishing offers a very good "back office" software package for small publishers. The program, *PUB123*, Version 3.0, is for Windows, but can run on Macs with emulation software. It handles order-taking, royalties, back orders, customer names and addresses, and all the other usual record-keeping chores for a small publisher. The program costs $199. I have purchased it, and use it for all the order-taking and royalty-tracking for FoxAcre Press. Adams-Blake also offers multi-user variants, for $199 per additional user. A downloadable free demo is available.

Fat Boys Software sells *Myrlyn*, a Windows-based program with roughly the same specs as PUB123. Myrlyn sells for $349, but membership in certain small-publisher's organizations gets you a $100 discount. To keep things in perspective, the price difference between Myrlyn and PUB123 is about the cost of one or two laser toner cartridges, and we're talking about the program you'll use to run your whole business. Don't be penny-wise and pound foolish. Take a good hard look at both of these programs, and make the price difference a low-priority consideration.

Moving up to where maybe the price point is a legitimate issue, we come to *Publisher's Assistant* 4.2d, a program formerly known as The Publisher's Information and Invoice Generating System, or PIIGS. *Upper Access Books* makes this program. What they call the "Sonnet" version for smaller operations is available for $495. Other versions would probably be more appropriate for a larger operation than PUB123 or Myrln. A more complex single-user "Lyric" version is $1,495, and a multi-user "Epic" network version is $2,245. Limited-function demos are available if you register at the site. One selling point for Publisher's Assistant is that you can start with the smaller version, and then upgrade later on without being forced to convert your data. As we'll see later in the chapter, this can be important.

Upper Access also offers the sort of fulfillment service discussed in the previous section. They will stock your books, take orders over their toll-free number, and ship them for you. See their web site for more information. Many companies offer similar services. As noted, *Para Publishing* lists lots of information about fulfillment services on their web site.

An outfit called *The Cat's Pajamas* sells a program by the same name intended for larger publishing operations. A smaller version of the program, called MiniCat, sells for $7,500 for a site license. It will run on Windows. The site licensing system means that multiple users on the network can use the system without need to pay for multiple licenses. A free demo CD-ROM is available.

Acumen, Inc. makes Acumen4 sofware. At last report, the cost for a single-user license was $6,950, and a three-user license cost $12,950. Acumen will sell more licenses, up to fifty or more, for the price difference between the existing number of licenses and the additional number of licenses the user wants. Various additional program modules to perform specialized tasks are available. It will run under various versions of Windows and on a Macintosh. A single user system should have 40 megabytes of RAM and at least a 1 gigabyte hard-drive. The network version can be run on a mix of Macs and Windows machines. A free demo and information package is available.

As best I can tell, PUB123, alongside the plain old Quicken check-balancing program, is as much bookkeeping software as my business ambitions are ever likely to need. I might well have gone with Myrlyn—but I found PUB123 first. Maybe someday I'll need to move up to Publisher's Assistant, but even if I did, I probably I wouldn't need most of what it can do for a good long time. Acumen4 and MiniCat are intended for a much larger small press than what mine is going to be for a while.

In any event, $13,000, or even $199, is far too much to spend for fulfillment software when what you need for the moment is a stack of three-by-five cards on which you can tick off the number of sales per book, and another stack of cards for writing down addresses. Why buy more than what you need?

However, the operative phrase there is "for the moment." If you hope and expect your business to grow, and hope and expect to have multiple titles and frequent sales to a long list of customers, doing the book-keeping and record-keeping by hand could rapidly get out of control and get you in big trouble.

Another solution would be to set up your own computerized database, somewhere between index cards and a full-blown fulfillment program in terms of sophistication. Of course, that would require you or someone else to set up a proper computerized database, which is no easy task for a beginner. Another possibility would be to see what customization is possible with the various standard book-keeping programs. Maybe someone else doing book-on-demand has set up a system of macros or templates or whatever for one of the $100 off-the-shelf bookkeeping programs, and would be willing to sell or give them to you.

One other note, that applies either to going from a paper or computerized homegrown solution to a commercial publisher's fulfillment system, or from a smaller-scale fulfillment system like PUB123, to a larger one, like Acumen4. Transferring your existing data into a new system, and entering it accurately and reliably, is a nontrivial task.

Obviously, typing in all the info off handwritten cards will be a tedious task, but don't assume that shifting from one computerized system to another will be any easier. I once worked on a small newsletter that converted from brand X to brand Y fulfillment software. Making the changeover, which was supposed to smooth and automatic, was a real headache. Crashes and scrambled data were the order of the day. (It was a very good and useful thing to have backed up the whole system first.)

There's another side to this coin: the cumulative investment of time and effort involved, in, say, a year's worth of keying in orders gets to be a serious issue. It can rapidly get to the point where the effort required to transfer all that data to a new system is so daunting that the idea of facing such a whacking dull and long task becomes the strongest argument against upgrading to a newer system. You're not just married to your current system:

you've got so much invested in it that you can't afford the divorce.

The people who sell the systems at least try to deal with this issue. Acumen4, for example, offers a program for converting your Cat's Pajamas data to their format. Adams-Blake offers a conversion service, and no doubt you can arrange for conversion with least some of the other programs. But I wouldn't count on such things working on the first try. (*And run a back-up first!*)

The best way to avoid such problems is to plan ahead, so that you will have the number-crunching capacity you need down the road. In other words, when you do set up your record-keeping and fulfillment system, think not only about the system you need today, but look ahead for the one you will need tomorrow.

7. Marketing

Marketing is probably the hardest single job in small-press publishing. You will almost certainly have to do for yourself, unless you are willing to pay out some pretty good-sized fees to a promotion company or public relations firm. There are lots of guides to promotion generally, and book promotion specifically, Visit John Kremer's website at www.bookmarketing.com for a lot of information—and a lot of chances to buy his books and reports. I am very definitely no expert on the subject of selling books. However, I can offer a few pearls of wisdom, gleaned from what I've read, and what I've learned from *trying* to sell books.

(1) Bookstores are often not the best place to sell self-published books, for reasons that have as much to do with how bookstores buy books as how they sell them. They usually buy through distributors or wholesalers, and won't be much interested in doing the paperwork to buy three copies of a book by a local author who wanders in. Nor are you going to do much business, even if you convince three bookstores in town to stock your title. They'll each take three copies, if you're lucky, and then probably never re-order. No one retires on the proceeds from selling nine copies at wholesale.

(2) Non-fiction is going to have a better shot than fiction, for two reasons.

First, the information is going to be the selling point. You bought *this* book because it's about book-on-demand printing, not because I wrote it. If you wrote a book about architecture, a bookstore that feels it needs more titles on that subject will buy copies.

In fiction, the writer's reputation is the selling point, or else the book sells because of something on the cover that inspires the reader to buy. If you wrote a novel about architects, it will help a lot if you've written six other successful similar novels and so have an established name—and it will hurt quite a bit if you don't. Secondly, the bookstore is going to know where to *put* a nonfiction architecture book—in the nice, clearly defined, architecture section. A novel just gets thrown into the huge fiction section, sorted by the author's last name, where it promptly vanishes from sight along with all the other titles by unknowns.

(3) Usually, your better shot will be to sell through stores or websites or catalogs connected to the subject of the book. Hardware stores for books on woodworking. Fishing store and bait shops for fishing books. Camera stores for books on photography. Computer stores for books on programming. People will go to those places looking for information on the subject in question. Even better, your book will be one of twenty or fifty, not one of five thousand. Relatively speaking, your title will be far more visible.

(4) A website helps, but not all that much. It's a billboard, and a place to send people who want more information. But there are so damned many websites out there that you can count on yours getting lost in the shuffle. A website is *not* a no-effort promotional device. You'll have to work at getting people to visit. Remember too that getting people to your website is not an end in itself. You want people to go there, and *then* buy

what you're selling. I'm not going to get started on the ins and out of selling off a web-page. That's a whole other book—and one I don't feel like writing.

(5) More is better. If someone really likes what you've sold, you should be in a position to sell them more. You can't do repeat business to retail customers if you only have one title to sell. Offer more books by the same author, or on the same subject.

(6) Keep at it. Marketing is an on-going job, not a one-time effort.

***

That brings us to the close of the main text of this book. Now you've got the basic information you need to design, print, bind, and sell your own books.

Go have fun with it.

Appendix One
Book-On-Demand Printing
As Appropriate Technology

In Chapter One I discussed several possible uses for book-on-demand printing. I mentioned self-publishing for profit, putting together a book of family remembrances, making up a church cookbook, creating parts catalogs or printing guides to museums, or other applications wherein demand is likely to be steady but slow, and/or where frequent revisions might be necessary. There are others. A high school could easily publish a book of student-written stories. A bookstore could put out its own line of private-label books.

But another possibility just might be the best use of all for these techniques, and I offer it here not only because it might change many lives for the better, but because it could get your imagination running, and get you thinking of new and better uses for books made in small batches.

The appropriate-technology movement is a big part of development projects in poorer countries around the world. The core concept of appropriate-technology is very simple: find machinery and ways of doing things that suit the needs and resources of the people that will use them.

In practice, this often comes down to choosing machines that are cheap to buy and use, rugged, reliable, and easy to replace or repair, rather than machines that do more, but require more money, time, and expense to maintain or operate. This does not necessarily mean rejecting new advances and new ideas. It just means choosing the new ideas that make sense.

Better to give a farmer a high-tech, high-strength, horse-drawn plough made from space-age materials that never needs sharpening, a plough that's scientifically designed for smoother pulling, rather than a tractor that will need expensive imported parts and fuel, and a trained technician to repair it. The vastly improved horse-drawn plough will still be in use years after the tractor is a rusting, abandoned hulk that died because there weren't any new four-dollar spark plugs for a thousand kilometers.

Appropriate technology is often a mix of the advanced and the very simple: for instance, a water pump powered by high-efficiency solar-power cells, so as to eliminate the need for stringing electricity out to the well site. It is also often a question of scale and decentralization: building seven or eight small bridges, some for foot traffic, others for wheeled vehicles, each designed and sited to best suit the local population, rather than one huge centralized bridge given over all but completely to vehicles, and inconveniently far from outlying communities.

When looked at from the right perspective, book-on-demand publishing is very much a mix of advanced and simple, and very much a scalable enterprise. Once the pages and covers of a book come out of the laser printer (or copier or digital duplicator, or whatever is used to produce the printed pages) the hand-assembly techniques for binding and trimming the books are all very low-tech and simple to set up. They do not require computers or ten-ton presses, but a person (or ten people, or twenty), a pot of glue, and a paper cutter.

Distributing the typesetting is the high-tech part of the job, and that technology is really pretty cheap, and getting cheaper. There are lots of ways to get the type to where the books are being made. You could just have a laser printer at each locale. Alternately, a central laser printer could produce a set of copy-ready pages of each book. You'd then have a digital duplicator (a sort of high-tech, high-speed version of the mimeo machines of old) or a photocopier at each local print site, ready to print out page sets. There would be lots of ways to scale the operation up or down as needed.

The covers would not need to be as complex or ornate as on commercially published titles, but both the covers and books would have to be extremely durable.

Book-on-demand printing could be a way for a small village

to establish a local publishing operation—attached to a school, perhaps. Books (and newsletters, etc.) could be published in the local dialect, and text books could be printed to order.

How-to books on any number of subjects could be stored in the computer (or as camera-ready physical pages) and printed as needed. Literacy would become a skill with immediate practicality, rather than as something theoretically desirable.

Obviously, the idea requires at least one person who understands the machines, a moderately reliable source of electricity, a distribution system that can deliver raw materials more or less on time and at a price that is more or less acceptable, et cetera. There are places in the world that just can't meet those conditions. But there are lots of places in the world that could meet them, places where books are scarce and expensive, if they exist at all.

Just as obviously, there are problems such as bootlegging and copyright violation, local censorship laws, and so on. Such problems will never be completely solved, but, over the years, publishers using conventional printing techniques have found ways to handle such headaches reasonable well. The book-on-demand printer would do the same.

Book-on-demand printing is also a project that's right-sized from the donor/capitalizer perspective: a modest-sized church or school in the United States, Canada, or Europe could easily raise enough money and/or scrounge enough used hardware to equip a third-world printing plant. As we will see in the next appendix, on *Sample Equipment Lists*, you can buy the hardware needed for a pretty slick book-on-demand operation for about the price of a good used car.

Nor is there any reason someone couldn't make a buck off appropriate-technology book-on-demand. A determined local entrepreneur could establish a publishing house with remarkably little capital, and run it very cheaply. He or she could do subsistence publishing, if you will.

There are any number of problems, and certainly there are lots of places where the idea would be impractical. But it seems to me that, if planned properly, book-on-demand printing of the sort discussed in this book could bring usefully large supplies of books to many places that have never had any such thing. That could make a lot of difference to a lot of people in the world.

Appendix Two
Sample Equipment Lists

General Equipment Requirements

It is possible to make books with very little equipment, but you can't expect to produce hundreds of books a day without hardware. In the sections that follow, we'll take a look a various possible equipment choices, and what they mean in terms of capacity and cost.

Computer and software pricing are extremely volatile, and of course some would-be book-on-demand operators will already have sophisticated setups, while others will start from scratch. Because so many will already have some computer equipment, and because prices are all over the place, prices for computer hardware are left blank in the following lists. Used computer hardware is often a bargain, by the way.

Cumulative Start-up

These three lists assume you start very small, and gradually work your way up to a more elaborate operation. You could even get by without buying some of the hardware I list as *essential*. You could eliminate the need for a computer or a printer by typing everything up by hand and having it photocopied.

But if your time is worth anything at all, the waste of time and effort required to get by without the *essential* list, and maybe even without the *highly-recommended*, list will quickly end up being penny-wise and pound-foolish. Book-on-demand is a manufacturing process, and you need the tools to do the job.

At a rough estimate, the *essential* list will do for the occasional weekend hobbyist who is only going to do a handful of books now and again. The *highly recommended* list is for those who move on to do more serious and on-going work. The *nice to have* list is for those with the money to spend, and/or those with

Essential Equipment	
fastest computer you can afford	—?—
capable word-processing program	—?—
laser printer capable of at least 600 dots per inch (DPI) and 12 pages per minute, (PPM) but faster if possible	$500
hand-operated guillotine paper cutter	800
"office" thermal binder	300
(or) high power glue gun and an electric griddle (total below assumes you choose this option)	100
misc. small tools (rulers, knives, caliper, paint brushes etc.)	50
initial supplies: paper, adhesives, printer toner, cover stock, etc.	200
Total	$1,650

Highly Recommended	
professional page layout software	$500
pro design or draw software	500
paper scoring machine	75
padding press	100
duplexing PostScript laser printer	1,000
color ink jet printer	500
Clickbook or other bookleting software	50
Total	$2725

Nice to Have	
basic page imposition software	$300
commercial thermal binder	6,000
high-speed duplexing laser printer	2,400
Total	$8,700

a reasonable expectation of producing a lot of books, enough books to justify a comparatively big investment. (But we'll be spending a lot more money in a page or two!)

As noted, the list is cumulative. The *nice to have* list assumes you already have the *essential* and *highly recommended* items. Taken literally, that means you end up with three printers, but that might well be what happens as you upgrade and yet retain old gear. Prices listed are my best estimate of current street prices for good-quality new hardware. While you can always pay more for fancier new equipment, on the other hand you can often find new or used hardware for less.

Cumulative list	*cost*	*cumulative*
essential	$1,650	$1,650
highly recc'd	3,325	4,975
nice to have	$8,700	$13,975

Author's Actual Equipment List

Let's compare this with a rough ballpark estimate of what I've spent on hardware over the last eight or ten years. And, by the way, the following list leaves off all the hardware I have since filed under "M" for mistake. (If I was being very kind to myself, I'd say I'd spend about another $3,000 on hardware that I have

Author's Equipment List	
computer system	$2,300
design and layout software	1,000
Lexmark T614nl duplexing laser printer	1,800
HP 1120ce color ink-jet printer	500
electric guillotine paper cutter (used)	450
Bind-Fast 5c binding machine (used)	2,250
paper scoring machine / rotary cutter	75
misc. small tools and machines	700
Total	$9,075

since retired, or that just didn't really work properly.) As you can see, I got lucky and found the two most expensive items, the binder and the paper cutter, used, and at good prices. Of course, it took a lot of digging around to find them. Be prepared to work hard at making your own luck.

Just to put these price lists in some sort of perspective, consider that the totals on these lists are more or less comparable to buying various sorts of used cars.

If you want a car, it's *essential* to have at least an old clunker that will usually get you there, more or less ($1,650), while it's *highly recommended* that you get a somewhat newer and better used car that will be faster, more reliable, and more efficient ($4,975). Even if it's not essential, it would be *nice to have* a late-model job, just a few years old, that's got lots of style and speed and features ($13,975). Of course, there's always more you can spend, as we'll see in a moment, when we spend enough to buy a whole fleet of new cars.

Needless to say, there are plenty of other things to spend money on. However, these lists should give you some idea of what you should expect.

But let me emphasize once again: do your own research. *Don't* rely on these figures in your planning. Prices change, machinery is introduced and then discontinued, and sometimes there just aren't any bargains out there, no matter how hard you search.

One other thing—the wonderful world of Ebay, and online auctions and online sales of used equipment, all started up *after* I had most of my hardware in place. Probably I could get a better deal, and have a better selection of items to choose from, and it would be easier to find them, if I were shopping today. On the other hand, the number of potential buyers for a given piece of equipment has gone up, so perhaps the odds are against finding too many dirt-cheap bargains. And do remember to factor in the cost and hassle factor of shipping big heavy machinery around the landscape.

Also remember these are not lists of everything you'll need to do business. For example, they don't include such things as mail-room supplies if you go into the mail-order business. And your choices might not be my choices. You might well decide to go with a new electric paper cutter, and that could easily run you $3,000, or even more. Factor a few such nasty surprises into your budget.

One very bad way to save money is by bootlegging software. When you start printing books, you are going into the intellectual property business. Stealing from others in the same line of work is no way to begin. Pay for what you use, even if it is expensive.

PageMaker, InDesign, FrameMaker, and Quark are the big desktop publisher packages, and they go for about $500 to $800 new. *Illustrator, Photoshop,* and similar graphics programs sell for about the same price.

That's the bad news. The slightly better news is that there are often deals on upgrades and so forth. Assuming the licencing agreements allowed it, you could, for example, buy an older version of a software package (say, on Ebay) and then, if you needed the newer features, pay the upgrade price for the current version.

The other good news is that, these days, it's hard to buy a computer or printer that doesn't come bundled with all sorts of software. An advanced word-processor will certainly be enough to get you started on page layout, and probably such a program came with your computer. There are various low-end DTP packages, in the $50 to $100 range, such as Serif's *PagePlus*, and Microsoft *Publisher,* that will also at least get you started, and might be all you'll ever need.

Equipment for Larger-Scale Operations

I started out my research on book-on-demand with the intent of setting up a table-top operation, and that was the primary focus of my research. However, I have learned enough to at least give a basic idea of what would be involved in setting up various sorts of larger-scale operation. The comparisons below, however approximate, might be of some interest. Prices on the high-end equipment are rough estimates at best. Prices are for new equipment. Note that these lists are strictly the equipment to produce books, not design or lay them out. I am assuming anyone spending this much money has already invested in some basic business equipment, such as an office computer system.

Furthermore, I have not broken out separate prices for computers for some of the large-scale system, as the computers for them (traveling under different names such as "front end" or "controller") are in essence bundled into the printer. I should also note that some of these prices are a few years old. As these lists are intended as a quick snapshot, not a definitive list, and as and prices have been reasonably stable, and as you can often find discounts anyway, I have not bothered to update them. Some of the items below will now be last year's model. The new models will do more, but cost about the same. Alternately, you ought to be able to pick up last year's model at a reduced price.

I should also note that all the printers discussed in the large-scale lists below are Xerox products. There are plenty of other printer manufacturers out there, but to simplify matters, I compiled just one set of prices. The competing products from IBM,

Océ, Ricoh, Canon, Lanier, and so on are roughly comparable in price.

As will be seen, the equipment for a large-scale setup can easily cost ten or fifteen times as much as my table-top operation, while providing a book-printing capacity that isn't that much greater (on paper, at least). Is it worth getting a mega-setup? I have not really decided how the bang-for-the-buck calculation works out. The case could be put forward that it would make more sense to buy ten of the table-top setups for $150,000, rather than one entry-level large-scale print line for $147,000. Maybe, or maybe not.

The printers for the large-scale operations are not just faster than the table-top system. They are also heavier-duty, and designed to print more pages. A $2,500 printer might be rated to do 30,000 pages a month. A $100,000 printer will be rated to do several million pages. The smaller printer will just plain wear out if subjected to the day-in, day-out use a bigger printer would take in stride. There is also the question of consumables. I do not have any numbers on what toner costs for a DocuTech, but I'd be willing to bet that it comes to far less than half of the 1.7 cents a page for my duplexing printer. Do 30,000 pages at 1.7 cents a sheet, and you've spent $510. Do a million pages, and you've spent $17,000. If the DocuTech toner cost .085¢ (that is, $0.0085) a page, you'd save $8,500. As a last argument, but without getting bogged down in detail, suffice to say the large-scale systems can also do various tricks the small-scale system can't. You might need a feature you just can't get in a smaller printer. The argument could go either way, depending on the application in question.

You can make similar arguments regarding the binding machines and cutters. On paper, they aren't that much faster or more powerful. However, they will last longer, and can produce better-quality books, and likely have features that the smaller units don't.

Once again, these tables list nothing but the equipment needed to do the basic work of printing books. There are plenty of other expenses and capital outlays, from buying postage to paying rent, that are not reflected here.

I start out with a basic table-top operation to provide a basis of comparison. This is a start-from-scratch list, rather than a cumulative one.

Table-top	
Fast computer and software	$2,000
32 ppm duplexing laser printer	2,000
Wide-carriage heavy-duty ink-jet printer	1,000
18" electric "guillotine" paper cutter	3,000
Bind-Fast 5c perfect binding machine	5,000
Misc supplies and equipment	1,000
Total	$14,000

I'd estimate the realistic maximum daily output of this table-top setup at approximately 50-100 books a day. The limiting factor would be the speed of the laser printer. Most of the time, the other machines are going to stand idle, waiting for the pages to be printed. Buying a faster printer or additional printers would significantly increase capacity. Note that most of the manufacturing processes require labor-intensive work.

Compare this to the large-scale start-up on the next page. Maximum realistic daily output of this system would, at a guess, be about 500 to 1,000 books a day (but I have a hunch that would be really pushing it). Because most processes will be far more automated that with the table-top system, this setup would require approximately same amount of labor to operate, excluding maintenance service to machines. In other words, one person could pretty much do all the work with either system. However, the machines for this large scale system would likely be fast

Large-scale	
DocuTech 6180 w/ front-end and options	$400,000
DocuColor 40 color printer w/front end	200,000
Automated programmable paper cutter	20,000
Standard BQ-140 or similar perfect binder	12,000
Case binding line	62,500
One-sided laminator	7,000
Signature sewing machine	190,000
misc additional equipment	20,000
Total	$911,500

enough that the machines would spend a lot of time waiting for the operator. Labor would be the limiting factor. It should also be noted that, with the above equipment, one could do both soft-cover and hardcover (that is, case-bound) books.

And if $900,000 sounds like a lot, bear in mind that it is always possible to spend more money. The 400 page-per-minute on-demand printing-and-binding line that IBM Infoprint displayed at various trade show cost about three million dollars.

“Entry-Level” Large-Scale

A gentleman from Xerox who didn’t wish to speak for attribution gave me his take on what it would take to set up an “entry-level” large-scale book-on-demand setup.

He emphasized that he was quoting what were called “non-negotiated” prices, with the strong suggestion that negotiation might help a lot. Equipment leasing is also possible. And, of course, used equipment is always available.

Note: this budget-priced setup assumes that you will be us-

Entry-Level Large-Scale	
DocuPrint 65 PPM printer	$50,000
Digipath front end with production software for above	25,000
Xerox DocuColor 5750 (aka Regal) 400 DPI color printer/copier	19,500
Splash Front-end controller for 5750	28,000
Standard BQ-140 perfect binder	12,000
Non-programmable paper cutter	3,000
Other equipment and supplies	10,000
total	$147,500

ing cover stock that won't require lamination. The list therefore omits the $7,000 for the one-sided laminator.

My gut reaction is that the above "entry-level" large scale system comes close to hitting the sweet spot between cost and productivity. This is equipment designed for fast, heavy-duty production. It's expensive, and certainly beyond the means of a home operation. But it's not beyond the dreams of a home operator who's thinking, if not big, at least bigger.

In other words, I wish *I* had toys like the ones in this list.

Appendix Three
Names and Numbers

Here is a list of useful names, addresses, and contact information for people, companies, and services connected with book-on-demand printing. Please bear in mind that addresses change frequently, especially email and Internet addresses.

I have done my best to update these addresses, and the list below reflects my best information as of August 2002. I will try and keep an updated list of contacts, with live web and email links, at http://www.foxacre.com/bookpage/bod-address.htm.

Some of these companies I have dealt with and liked. Some I have had no dealings with of any kind. Some I have found to be marginally competent. I include the less impressive companies and the unknown quantities because book-on-demand information can be hard to come by. Sometimes you have to take what you can get.

This is not a comprehensive list. There are no doubt lots of good companies that I have failed to include. If you have any notes, comments, corrections or additions, please send them to me care of FoxAcre Press at info@foxacre.com.

3M
3M Product Info Center
3M Center Building 304-1-01
St. Paul MN 55144-1000
(800) 3M-HELPS(651) 737-6501
fax: (800) 713-6329
innovation@mmm.com
www.mmm.com

A Round Table Solution (ARTS)
Level 10, 114 Albert Road
South Melbourne Victoria 3205
Australia
(61) 3 9686 5522
fax: (61) 2 9475 0464
PDF software. See Binary Thing.

Acumen Inc.
1596 Pacheco
Suite 203
Santa Fe New Mexico 87505
(505) 983-6463
fax: (505) 988-2580
www.acumenbook.com
Makers of Acumen4 publisher's bookkeeping software.

Adams-Blake Publishing
8041 Sierra Street
Fair Oaks CA 95628
voice/fax (916) 962-9296 orders
(800)368-2326

www.adams-blake.com
Makers of PUB123 publisher's bookkeeping software.

Adobe Systems Incorporated
411 First Street S.
Seattle Washington 98104
phone: (800) 833-6687(408) 986-6520 (tech support)
www.adobe.com

AHT (Advanced Hi-Tech)
1990 E. Grand Avenue
El Segundo California 90245
(800) 538-8287 or (888) AHT-AHT7
fax: (310) 615-1871
www.aht.com
distributors of binding equipment

Allen, Roger MacBride
info@foxacre.com
www.rmallen.net
Author of this book.

Banner American Products Inc.
42381 Rio Nedo
Temecula CA 92585
(909) 296-9780

fax: (909) 296-9790
foliant@banam.com
www.banam.com
Distributors of Foliant laminators and other laminators and lamination supplies.

BarcodeWeb
www.barcodeweb.com
on-line barcode provider. Web site down. Out of business?

Binary Thing
Level 10 144 Albert Road
South Melbourne Victoria 3025
Australia
(61) 3 9686 5522
info@binarything.com
www.binarything.com
Makers of various add-on tools for Acrobat. See A Round Table Solution.

Blitzprint
#3 1235 64 Ave SE
Calgary Alberta T2H 2J7 Canada
(866) 479-3248
Canadian book-on-demand printer.

Blue Squirrel
170 West Election Drive Suite 125
Draper Utah 84020
(800) 403-0925(801) 523-1063
fax: (801) 523-1064
sales@bluesquirrel.com
www.bluesquirrel.com
Makers of the excellent Clickbook page-reordering software.

Book Automation Inc.
356 Ely Avenue
Norwalk CT 06854
(203) 853-7200
fax (203) 853-6809
www.bookautomation.com
sells Kristec automatic book sewer

Book Marketing Update/Open Horizons
PO Box 205
Fairfield Iowa 52556
(800)796-6130, (641) 472-6130
fax: 641-472-1560
John Kremer's business, promoting book marketing. Lots of information on the website on promotion, book printing, etc.

BookMakers International
6701B Lafayette Avenue
Riverdale Maryland 20737
(301) 927 7787
fax: (301) 927 7715
e-mailbookmkrs@aol.com
www.bookmakerscatalog.com
hand binding and hardcover binding supplies

Bookmasters
2541 Ashland Rd
Mansfield OH 44905
800-537-6727
www.bookmasters.com
Book printer. Also offers fulfillment and other services. Online bookstore at www.atlasbooks.com.

BookMobile
2402 University Ave W
Suite 206
Saint Paul, MN 55114
(651) 642-9241
fax: 651-603-9263
www.bookmobile.com
short-run and on-demand book printer.

Brodart Co.
P.O. Box 300
100 North Road Clinton County Industrial Park
McElhattan Pennsylvania 17748
(800) 233-8467
Fax: (800) 283-6087
www.brodart.com
binding supplies.

Brother International Corp.
100 Somerset Corporate Boulevard
Bridgewater NJ 08807-0911
(908) 704-1700
fax: (908)704-8235
www.brother.com
makers of Cool Laminator & many other products.

C.P. Bourg
New Bedford Industrial Park 73
Samuel Barnet Blvd
New Bedford MA 02745
(508) 998-2171
fax: (508)998-2391
www.cpbourg.com
Binding and cutting machines.

Coverbind Corporation
3200 Corporate Drive
Wilmington NC 28405
(800) 366-6060
fax: (910)799-3935 Office Thermal Binding systems.
www.coverbind.com

Digitz.net
270 King Street
Charleston SC 29401
(866) 308-6235fax: (843) 577-7506
fax: info@digitz.net
www.digitz.net
on-demand printing service. Formerly independent, now connected to Book Surge.

Dover Publications Inc.
31 East 2nd Street
Mineloa New York 11501
fax: publisher of reprint books
www.doverpublications.com
many on craft topic. Many books on bookbinding. Write for catalog.

Duplo USA Corporation
3050 South Dailmer Street
Santa Ana California 92705
(714) 752-8222
fax: (714) 752-8222

www.duplousa.com
U.S. subsidiary of Japanese binding equipment maker.

Electric Press Inc.
Reston VA
www.elpress.com
Offers "read before you buy" web service to publishers. Status of business very uncertain. Might or not might be going concern.

Epublishstore.com
www.epublishstore.com
Sell PDF manipulation software. Website links to many PDF sites.

Evans, Rupert
r-evans4@uiuc.edu
www.staff.uiuc.edu/~r-evans4
author of the extremely useful guide *Book on Demand Publishing*

ExactBind Corporation
3491 Pall Mall Drive Suite #103
Jacksonville Florida 32257
(904) 880-2206
exactbind@aol.com
A U.S. distributor of Maping's Fastbind binding system.

Factory Express
1720 Coulter St
Rio Rancho New Mexico 87144
800-399-2564505-892-9600
www.factory-express.com
Laminators, binding machines, paper cutters

Farrukh Systems
125 Shenley Road
Borenhamwood Hertfordshire WD6 1AG U.K.
0181-207-4249
fax: 0181-207-5992
fsupport@farrukh.co.uk
www.farrukh.co.uk
Makers of Imposition Publisher PostScript imposition software.

Fat Boys Software
201 Kenwood Meadows Drive
Raleigh North Carolina 27603
(919) 773-2080
www.myrlyn.com
makers of Myrlyn publisher's software.

Fibermark DSI
1 Canal Street
South Hadley Mass. 01075
(800) 843-1243(413) 533-0338
fax: (413) 532-4810
dsimarketing@fibermark.com
www.fibermarkdsi.com
(formerly Rexam DSI). Makers of ImagEase stock for paperback covers

FinePrint Software
16 Napier Lane
San Francisco CA 94133
fax: (775) 254-1923
sales@fineprint.com
www.fineprint.com
Makers of Fineprint print-reordering software and pdf generation software.

Flash Magazine
Riddle Pond Road
West Topsham Vermont 05086
(800) 252-2599
info@flashweb.com
www.flashweb.com
Internet-published magazine. Semi-inactive.

Foliant
www.foliant.com
laminators. Distributed by Banner American. As we go to press, web site is blank.

GBC Canada
49 Railside Road
Don Mills Ontario M3A 1BC
Canada
(800) 463-2545
www.gbccanada.com

GBC
General Binding Corporation
One GBC Plaza
Northbrook Illinois 60062
(800) 723-4000
www.gbc.com
Binding and laminating.

GF Murray Company
2863 Wall Street
2nd Floor
Vancouver British Columbia
V5K1B1 CANADA
(604) 253-1046
fax: (604) 253-1056
www.gfmurray.com
Short-run printing & book-on-demand publisher

GhostScript/Aladdin
Box 60264
Palo Alto California 94306
(415) 322-0103
www.cs.wisc.edu/~ghost/index.html
GhostScript and Ghostview are available as downloads.

Gigabooks
P.O. Box 90674
Honolulu HI 96835
www.gigabooks.net
sells how-to books and equipment for hand-binding.

Hewlett-Packard
Santa Clara California 92122
(800) 752-0900
fax: (208) 323-2551
www.hp.com

Hollander's
407 North Fifth Avenue
Ann Arbour Michigan 48104
(734) 741-7531
fax: (734) 741-7580
hollanders@earthlink.net
www.hollanders.com
hand-binding and hardcover binding supplies.

IBM Printing Systems
(800) 358-6661
fax: (404) 238-1234
www.printers.ibm.com
Makers of IBM printers. See Lexmark

ICG /Holliston
Hwy. 11-W
Holliston Mills Road
Church Hill Tennessee 37642
(800) 325-0351(423) 357-6141
fax: (423) 357-8840
custsvr@icg-online.com www.icg-online.com
Makers of linen and other stocks for case (hardcover) bindings.

Imation Enterprises Corporation
1 Imation Place
Oakdale MN 55128-3414
(800) 826-4886
fax: (800) 552-2039
www.imation.com
info@imation.com

Import-Graphics Inc
267 Fifth Avenue Suite 610
New York NY 10016
212.532.1988
U.S representative for Planax, a binding equipment manufacturer.

Independent Publisher
see Jenkins Group.

Instabook
Gainesville Florida
fax: (352) 371-1154
ibc@insta-bookcorporation.com
www.instabook-corporation.com
Start-up company that is just starting to market all-in-one book-making machines.

Intelli Innovations Inc.
923 Griffis Street
Cary NC 27511
(919) 468-0340
www.intellisw.com
Makers of barcode image software for Mac and Windows.

IUniverse.com
www.iuniverse.com
book-on-demand publisher and publishing service provider.

J. Hewit & Sons
Unit 28 Park Royal Metro Centre
Britannia Way
LONDON NW10 7PR
United Kingdom
01811 965-5377
fax: 0181 453-0414
www.hewit.com
British supplier of leather and bookbinding supplies. US distributors are Bookmakers International.

Jenkins Group
121 East Front Street 3rd Floor
Traverse City Michigan 49864
(616) 933-0445
fax: (616) 933-0448
jgbp@northlink.net
www.bookpublishing.com
Publishers of Independent Publisher. Note: Independent Publisher has shifted from print publication to on-line publication.

Jet Print Photo
(800)323-4193
www.jpphotodirect.com
A division of International Paper. Makes coated stock suitable for printing covers in ink-jet printers.

Kremer, John
see Book Marketing Update Open Horizons

Lancaster, Don (Synergetics)
3860 West First Street
PO Box 809-W
Thatcher AZ 85552
(520) 428-4073
Fax: (520) 428-6630
don@tinaja.com www.tinaja.com
Self-styled Book on Demand and PostScript guru.

Laser World
www.laserworld.net
toner cartridge refiller.

LBS (Library Binding Service)
2134 East Grand Avenue
P.O. Box 141
Des Moines Iowa 50305-1413
(800) 247-5323
fax: (800) 262-4091
lbs@lbsbind.com
www.lbsbind.com
Binding supplies.

Legend Communications Inc.
54 Rosedale Avenue
West Brampton
Ontario L6X 1K1 Canada
(905) 450-1010
fax: (905) 455-9702
www.legendcomm.com
Makers of Double-Up Page Imposition Software. Phone disconnected. Out of business?

Leonard's Distributing
4125 Prospect Drive
Carmichael California 95608
(916) 967-6401
Distributed Unibind and other lines. Out of business?

Lexmark International
740 New Circle Road N.W.
Lexington Kentucky 40522-2876
(800) 539-6275, (606) 232-2000
www.lexmark.com
Makers of Lexmark printers. See IBM printers.

Lightning Source
1136 Heil Quaker Blvd.
La Vergne TN 37086
(615) 287-5815
www.lightningsource.com
formerly Lightning Print. on-demand printing service owned by Ingram, a book distributor.

Maping
Atomitie 5 F
HELSINKI 00370 Finland
+358 9 562 6022
fax: +358 9 562 6215
www.maping.com
Finnish company that manufactures Fastbind Binders.

Marsh Technologies, Inc
761 Spirit of St. Louis Blvd
Chesterfield MO 63005
(636) 530-0100
fax: (636) 530-1232
jmarsh@marshtechinc.com
www.marshtechinc.com
Maker of "black box" book print & bind system.

Martin-Yale
251 Webcor Avenue
Wabash Indiana 46992
(219) 563-0621
www.martinyale.com
paper cutters.

Microsoft
www.microsoft.com
This is some sort of software company. A few people have done business with them.

Minolta-QMS Inc.
One Magnum Pass
Mobile Alabama 36618
(800) 523-2696(334) 633-4366
info@minolta-qms.com
www.minolta-qms.com
makers of well-regarded high-end printers.

Mitchell Graphics
2363 Mitchell Park Drive
Petsokey Michigan 49220
(800) 583-9403
mgi@mitchellgraphics.com
www.mitchellgraphics.com
Short-run specializing in full color work. Focuses on postcards for promotion, but might be a source for cover printing.

Mitsubishi Imaging
555 Theodore Fremd Avenue
Rye New York 10580
(888) 258-7919
fax: (914) 921-0995
digital@mitsubishiimaging.com
www.mitsubishi-imaging.com
Paper stocks suitable for ink-jet covers.

National Binding
3222 Roman Street
Metairie Louisiana 70001
(800) 878-2463
fax: (504) 836-2463
Unibind distributor.

National Toner Recycling and Supply
29 Harbor Avenue
Norwalk Connecticut 06850
(800) 676-0749
fax: (203) 853-1258
www.nationaltoner.com
Laser toner cartridge reloader.

North East Printing Machinery
http://www.nepminc.com/
Has on-line listings of printing equipment for sale.

ODM (On Demand Machinery)
150 Broadway
Elizabeth New Jersey 07206
(908) 351-7137
fax: (908) 351-7156
www.agamachinery.com
Makes on-demand case binding machines. Also sells used equipment.

Office Zone
1142 West Flint Meadow Drive
PO Box 121
Kaysville UT 84037
800-543-5454801-927-3026
info@officezone.com
www.officezone.com
Good prices on binding machines, laminators, other equipment, and supplies.

On Demand Machine Corporation
P.O Box 8470
St. Louis Missouri 63132
(702) 995 9007
information@bookmachine
www.bookmachine.com
Maker of all-in-one book making machines. This is a entirely different company from the previous entry. Website information seems dated.

Paper Plus
(888) PAPERPLUS
www.paperplus.com
National chain of paper suppliers.

Para Publishing
PO Box 8206-240
Santa Barbara CA 93118-8206
(805) 968-7277
fax: (805)968-1379
info@parapublishing.com
www.parapublishing.com
Publisher of *The Self-Publishing Manual*. The web site contains lots of useful info.

PDF Snake
8345 NW 66th St, Suite 3168
Miami FL 33166-2626
www.pdfsnake.com
Acrobat PDF imposition software.

Peachpit Press
1085 Keith Avenue
Berkeley California 94708
publisher of books on desktop publishing
www.peachpit.com

Pendl Company
1825 B Dolphin Drive
Waukesha WI 53186
(800) 869-7973
pendl@pendl.com www.pendl.com
Toner cartridge recharger.

Perfect Systems
www.perfect-systems.com
See Marsh Technologies.

Planax
See Import-Graphics.

POD Wholesale
519 W. Lancaster Ave
Haverford PA 19041
(610) 520-2500
fax: (610) 519-0261
info@podwholesale.com
www.podwholesale.com
print-on-demand book printer.

Powis Parker
775 Heinz Avenue
Berkeley California 94710
(800) 321-BIND
fax: (510) 848-2463
www.powis.com
Binding machines.

Printer's Shopper
111 Press Lane
Chula Vista California 91910-1093
(800) 854-2911
fax: 1-800-482-8563
sales@printersshopper.com
www.printersshopper.com
A Division of NewCo Printing Supply. They offer a very useful catalog, geared toward the traditional print shop, but with many products useful for b.o.d.

Publisher's Marketing Association
627 Aviatiion Way
Manhattan Beach Ca. 90266-7107
(310) 372-2732
fax: (310) 374-3342
www.pma-online.org
Organization to support small press publishers.

Quill Corporation
P.O. Box 94080
Palatine Illinois 60094-4080
(800) 789-1331
www.quillcorp.com
Office supplies and equipment.

Quite Software
Carraig Thura, Lochawe
Near Dalmally PA33 1AF U.K.
011 44 20 8553 6574
fax: 011 44 1631 574088)
sales@quite.com www.quite.com
Makers of Quite Imposing Software.

R.R. Bowker
121 Chanion Road
New Providence NJ 07974
(908) 464-6800 www.bowker.com
ISBN and *Books In Print* publisher.
Call (908) 665-6770 for ISBN prefix application.

Recharger
4218 W. Charleston Blvd.
Las Vegas
Nevada 89102
info@rechargermag.com
www.rechargermag.com
A big glossy magazine and website dedicated to the laser toner cartridge reloading industry.

Replica Books
(908) 541-7391
www.replicabooks.com
on-demand printing service owned by Baker & Taylor, a book distributor.

RJ Communications
51 East 42nd Street
Suite 1202
New York New York 10017
800-754-7089
fax: 310-406-2363
www.booksjustbooks.com
short run & on-demand printer. Does full-color children's books.

Rosback Co. (F.P. Rosback)
125 Hawthorne Avenue
St. Joseph Michigan 49085
616-983-2582
fax: (616)983-2516
www.rosbackcompany.com
binding equipment.

Serif Inc.
PO Box 803
Nashua NH 03061-9885
(800) 557-3743
Sales@Serif.com www.serif.com
Makers of Page Plus Page Layout Software.

Small Publishers Association of North America
P.O. Box 1306
425 Cedar Street
Buena Vista CO 81211
719-395-4790
fax: (719) 395-8374
span@SPANnet.org
www.spannet.org

SNX
692 10th Street
Brooklyn NY 11215-4502
(800) 619-0299
snx@snx.com www.snx.com
on-line barcode provider & barcode software for Mac & Windows.

Spiral Binding Company
P.O Box 286
One Maltese Drive
Totowa New Jersey 07511
(800) 631-3572
fax: (201) 256-5981
www.spiralbinding.com
binding and laminating supplies.

Standard Finishing Corporation
10 Connector Road
Andover MA 01810
(800) 526-4774
fax: (978) 470-2771
www.sdmc.com
binding equipt. Standard Duplicating is another div. of the same co.

Standard Graphics Mid-Atlantic
3514 Lee Highway
Arlington Virginia 22207
(800) 723-2510
fax: (703) 524-2125
Distributors for Standard Finishing and other lines.

StickerMaker.com
1159 Leonard NW
Grand Rapids MI 49504
(616) 456 5585
fax: (616) 456 6083
info@stickermaker.com
www.stickermaker.com
Web sales name of KM Services & Supply, authorized distributor of Xyron laminators.

Talas
568 Broadway
New York New York 10012
(212) 219 0770
fax: (212) 219 0735
e-mailinfo@talasonline.com
www.talas-nyc.com
hand binding and hardcover binding supplies.

TechPool Software
2726 Loker Avenue West
Carlsbad CA 92008
800-925-6998
support@techpool.com
www.techpool.com
makers of Transvertor Pro PostScript manipulation software.

TeMPeR Productions
117 South 14th Street
Olean New York 14760
(716)373-9450
info@temperproductions.com
www.temperproductions.com
makers of fan-gluing presses and other binding equipment

The Cat's Pajamas, LLC
12559 Pulver Road
Burlington Washington 98233
(800) 827-2287
fax: (360) 707-5400
info@tcpj.com www.tcpj.com
Makers of The Cat's Pajamas and MiniCat publisher's bookkeeping software.

The Print Shopper
www.halcyon.com/shopper
Has on-line listings of printing equipment for sale.

Thomson-Shore
7300 W. Joy Road
Dexter Michigan 48130
(313) 426-3939
fax: (313) 426-6216
www.tshore.com
Highly-regarded short-run printer. Good information available at web site.

TPX Online
www.tpxonline.com
Has online listings of printing equipment for sale.

Trafford Publishing
2333 Government St Suite 6E
Victoria British Columbia V8T 4P4
CANADA
(888) 232-4444, (250) 383-6864
fax: (250) 383-6804
webmaster@trafford.com
www.trafford.com
book-on-demand publisher. Has office in US at: 5804 Jolly Roger Court, New Bern, NC, 28560-9767

Unibind Inc
11810 Wills Road
Suite 100
Alpharetta GA 30004
(770)674-6000
fax: (770) 674 6007
mailer@unibindna.com
www.unibind.com

Makers of office binding equipment that's not really appropriate for book-on-demand.

Unibind Western (USA)
5960 South 318th
Auburn Washington 98001-0823
(253) 735-6334
fax: (253) 735-6304
Distributor for Unibind.

Upper Access Inc.
1 Upper Access Road
P.O. Box 457
Hinesburg Vermont 05461-0457
800-310-8320 (orders only)(802) 482-2988
fax: (802) 482-3125
info@upperaccess.com
www.upperaccess.com
Makers of Publisher's Assistant publisher's bookkeeping software. Also provides fulfillments and book promotions services.

Van Son Holland Ink
92 Union Street
Mineola New York 11501
(800) 645-4182
fax: (516) 294-8608
www.vansonink.com
ink-jet refill systems and supplies.

Vaughan Printing
411 Cowan Street
Nashville Tennessee 37207
(615) 256-2244
fax: (615) 259-4576
bookprint@aol.com
www.vaughnprinting.com
Book printer that I've heard good things about. As of this writing, website is merely a blank page saying "under construction."

Wisdom Adhesives
10275 Pacific Avenue
Franklin Park Illinois 60131-1625
(847) 678-7750
fax: (847) 678-7793
info@wisdomadhesives.com
www.wisdomadhesives.com
Adhesives for b.o.d.

Xerox
www.xerox.com
(800) 34-XEROX in Canada (800)-ASK-XEROX
www.xerox.com

Xlibris
(888) 7XLIBRIS, (215) 923-4686
fax: (215) 923-4685
service@xlibris.com
www.xlibris.com
On-demand publishing service provider. Offers picture-book publication as well as text-only.

Xyron
15820 North 84th Street
Scottsdale Arizona 85260
(800)793-3523, (480)443-9419
fax: (480) 433-0118
www.xyron.com
makers of cold-lamination machines and supplies.

Index

Numbers

3M Corporation 62

A

Accubind Document Binding System 130
Acrobat 85
Acumen 193
Acumen4 publisher's software 193
Adams-Blake Publishing 192
adhesives
  for book binding 119
  for production binders 141
Adobe Magazine 11
Advance Reading Copies (ARCs) 14, 35
Allen, Roger MacBride 230
anvils, used 144
Appropriate Technology
  and book-on-demand 198

B

Barcodes 93
Baum 154
BaumBinder 300 154
Bibliobytes 28
Bind-Fast 5 146
Binding
  cold glue 117, 153
  lay-flat 118, 153, 155
Binding Machines
  floor-model 150
  small 124
  table-top 146
Binding Strip Machines 127
Black Box Printing Systems 159
book printing
  basic tasks of 34
  conventional 3
Book-On-Demand Printing As Appropriate Technology 198
Book-On-Demand Publishing, book by Rupert Evans 7, 9, 49, 106,112, 115
bookmarketing.com 195
Books
  full-color 181
  hardcover 180
  picture 180
  terms for types defined 106

Booksurge 183
Bourg Binder BB2000 152
Bourg Perfect Binder BB1000 149
Bowker 93. *See* R.R. Bowker Company
BQ-140 152
BQ-P6 131
Brodart library supplies catalog 63

C

C.P. Bourg 149, 152
CD-ROM 187
Clickbook page reordering software 82
CMYK Books 183
Cold Glue Binding 117
Cold-Glue Binding 153
Conglomco Enterprises 143
Colophon, 229
contact cement
  for book-binding 122
contract, book 35, 175
Conventional Publishers 169

cost per book 34

D

DB-250 148
DeScribe Word Processor 68
Design
  cover 92
  page 64
desktop publishing rograms.
  See Page-layout programs
Digital Printers 169
Digital Warehouses 178
Duplex Printing 77
duplexing 40
Duplo 148

E

E-Books
  proprietary-format 29
  web based 27
  websites related to 31
Ebay 42, 52.
Economics of Book-On-Demand Printing 2
Equipment for Printing and Paper Handling 40
Evans Do-It-Yourself Thermal Binding Systems 112
Evans, Rupert 7, 9, 11, 49, 53, 55, 112, 113, 115, 123, 230
Exactbind 132, 133

F

Fan-gluing 120
Fastback Spine Tape Binding System 127
Fastbind 132
Fat Boys Software 192
Fibermark DSI 99
Fields, W.C 189
Fineprint page reordering software 82
Flash Magazine 10
Foliant laminators 60
fonts 64
FoxAcre Press x, 7
FrameMaker 68
Fulfillment 190
Fulfillment Services 190
Fulfillment Software 190
Full-color books 181

G

GBC 60
Ghosts of the Internet 188
GhostScript 85
GhostView 85
Gigabooks 121
glue
see also adhesives
  for book binding 119
grant of rights 176
guillotine paper cutter 48

H

H530 Hard Cover Maker 136
Halfback 129
Hardcover Books 180
hardcover books, making 106, 136
Helvetica font 65

I

ideas, unbelievably stupid 188
Illustrator 68
ImagEase
  cover stock 99
Imation 51
Imprintbooks 183

Independent Publisher Magazine 11
InDesign 66
Instabook 160
International Standard Book Number 172
International Standard Book Number (ISBN) 25, 37, 172
  and barcodes 93

K

Kremer, John 195
Kristec Automatic Book Sewing Machine 158
Krylon UV resistant clear spray varnish 58

L

laminating, hand 63
Laminators 58
Lancaster, Don 9
Lay-Flat Binding Machines 153, 155
Layout
  cover 101
  pages 64
Lexmark 43
Lightning Source x, 93, 178, 179, 229
lying press and plough 50

M

manuscript
  unpublishable 13
  unpublished 16
  unsolicited
    odds of getting published 16
Maping 132
Marketing 36, 195
Milling 139
MiniCat publisher's software 193
money should always move toward the writer 18, 173
Myrlyn publisher's software 192

N

Neo-Vanity Presses 173
North East Printing Machinery 52
Notching 139
Novicki, Chet 122
Number Crunching 189

O

Office Thermal Binding Systems 108
On Demand Machine Corporation 160
On-Demand Case Binding Line 159
Otabind 155

P

padding 143
padding press 120
page description languages 41
Page Imposition 79, 85
Page-layout programs 66, 67
PageMaker 66
PagePlus 68
Paper
  cover stock 97
  curl 71
  custom sizes and avoiding cutting and imposition 84
  grain 69
  opacity 69

paper cutters 48
paper cutters, guillotine 48, 52
paper grain 57
PCL 41
PDF manipulation programs 90
PDF Snake imposition software 90
Penta-Bind 108
perceived value 28
Perfect Binder FII 155
Perfect Systems 160
Picture Books 180
PIIGS 192. See also Publisher's Assistant: publisher's software
Pillsbury© Brand® Breadtm Flour© 111
Planax 155
PostScript 41, 85
Powis Parker 127
Powis Parker Scoring Machine 56, 129
Printers
  book, digital 169
  book, short-run 169
printers
  "Windows-Only" 47
Printers, Ink-Jet 46
Printers, Laser
  Monochrome (black) 40
  color 45
Printing
  duplex 77
  Twin Two-up 75
  Two-up 71, 75
Printing, Covers
  doing it yourself 96
  using an outside printer 94
Private Publishing 25
Product Research 107
Production Printing 186
Prototyping 185
PUB123 publisher's software 192
Publication
  conventional 18
Publisher 68
  conventional 169
publishers
  as bankers 19
  as middlemen 21
  forgetting to pay authors 22
Publisher's Assistant publisher's software 192
Publisher's Software 190
Publishing
  Contract 175
publishing
  basic tasks of 21
publishing businesses
  start-up 188
Publishing Service Providers 27, 165, 171

Q

Quark 66
Quite Imposing imposition software 90
Quite Software 90

R

Recharger Magazine 44
Repair
  of paperback books 115
RepKover 155
reversion clause
  in book publishing contract 177
RJ Communications 183

RKM 200 155
Rosback 850 Series binding machines 151
Rosback Company 151
Roughening 139

S

saw brush for spine roughening 133
Scammers 173
Scoring 54
  and Maping's Fastbind system 133
scoring machines 55
Self-Publishing 25
Self-Publishing Manual, The, book by Dan Poynter 11, 26,37, 219
Side-Stapling
  as binding technique 121
Smythe-sewing 106
Software
  for page design 66
Sources for Information 9
Soviet Ministry of Agriculture 111
Standard Duplicating 130
Standard Finishing Systems 131, 146, 152
subsidy publishing. *See* Vanity Press
supply and demand in publishing for authors 22
Synergetics 9

T

Tektronic solid ink printer 45
TeMPeR Productions 120
The Cat's Pajamas publisher's software 193
Times font 64
toner cartridges
  cost of 43
  remanufactured and reloaded 43
  "killer" chips to prevent reloading 43
toner cost 42
TPX Online 52
Trashcanistan 146
true-believer syndrome xii
Two-Up Printing 71
typeface selection 65
Typography 64

U

Unibind 108
Upper Access Books 192

V

Vanity Press 23
Varnishes 58
Volkswagen Bug 146

W

Word, Microsoft 66
Word Pro, Lotus 66
Word processors 66
WordPerfect 66

X

Xerox 45
Xlibris 182
Xyron laminating machines 61

Z

ZIP drives 187

Colophon

I have printed many different versions of this book in many different ways. I update the text, and my procedures, frequently. This colophon therefore applies to this copy of the book, but not necessarily to any others.

Prior drafts of the book were written in the DeScribe 5.0 word processor, a good program which is now defunct. I have now imported all text in Microsoft Word 2000, even though I don't much like Word, because Pagemaker's import filter for Word picks up index marks and other features. The rough draft of the text was imported in PageMaker 7.01 running under Windows 98SE, and edited in PageMaker. Many of the illustrations in this edition, are based on the old DeScribe images. They have been brought into Adobe Illustrator 10 and reworked from there. EPS files generated by Illustrator were placed in the PageMaker layout files. The book is set in Times and Helvetica in various point sizes and styles, the only exceptions being the fonts in the FoxAcre logo. The cover file was likewise produce in PageMaker.

Some copies of the book are printed by Lightning Source, Inc (LSI). These copies have the Lightning Source logo on the back cover. Copies I print myself have a white star in a blue circle on the spine. Because of differences between print drivers, and lags in getting updates to LSI, there may be minor variations in text flow and text content between the two versions.

In order to generate the files for LSI, I "print" the final layout to a PostScript disk file. I then produce an Acrobat file from the PostScript file. (LSI prints from Acrobat and PageMaker files, as discussed in the text.)

For copies produced in-house, I print the pages from PageMaker or the PDF files, using Clickbook to produce twin-two-up print files, which I save and print from as needed.

I print the covers on a Hewlett-Packard Deskjet 1120ce, and cover the exterior of the covers with clear laminating sheets applied by hand. The pages I print on my Lexmark T614nl duplexing laser printer. I cut the pages down to half-letter size on my old 15-inch Ideal guillotine paper cutter. The book pages are bound into the covers using the Standard Bind-Fast 5, and the books are trimmed to final size on the guillotine.

A Note At The End

I should like to extend my thanks to Dr. Rupert Evans, who read over this book, caught several errors large and small, and made many useful suggestions. I can always learn more from him. If *you* know something I ought to learn, contact me at FoxAcre's email address, info@foxacre.com, with your thoughts and advice.

I can sum up what I have learned in the process of writing this book in a very few words: New materials and processes have made it easier and easier to make your own paperback books, and to make better and better books. For a relative modest investment, you can be off and running. So go get started.

And, in closing, I can tell you this: holding a book in your hands when you've just finished making it yourself is one of the more satisfying experiences around.

I hope you have it soon, and often.

RMA

About The Author

Roger MacBride Allen was born in Bridgeport, Connecticut on September 26, 1957. He graduated from Boston University in 1979 with a degree in journalism, and published his first novel in 1984. From that time to this, every work of science fiction that he has completed has been published. He has written eighteen novels to date, (three of which were New York *Times* bestsellers) and a modest number of short stories.

In 1994, he married Eleanore Fox, an officer in the U.S. Foreign Service. In March 1995, they moved to Brasilia, Brazil, where Eleanore worked at the embassy. In August, 1997, Eleanore's next assignment took them back to the United States. Their son, Matthew Thomas Allen, was born November 12, 1998. Though their home is in Takoma Park, Maryland, they will be on overseas assignment in Leipzig, Germany for three years starting in September 2003

www.ingramcontent.com/pod-product-compliance
Lightning Source LLC
Chambersburg PA
CBHW031925190326
41519CB00007B/413
* 9 7 8 0 9 7 0 9 7 1 1 8 0 *